AF687448

Cell and Molecular Biology Essentials

A Comprehensive Guide to Cellular Structures, Functions, and Molecular Mechanisms

Written by

Bolakale Aremu

Cell and Molecular Biology Essentials

A Comprehensive Guide to Cellular Structures, Functions, and Molecular Mechanisms

Published in the United States

ISBN: 979-8-3492-4281-6

Disclaimer

Although the authors and publisher have made every effort to ensure that the information in this book was correct at press time, the authors and publisher do not assume and hereby disclaim any liability to any party for any loss, damage, or disruption caused by errors or omissions, whether such errors or omissions result from negligence, accident, or any other cause. This book is not intended as a substitute for the medical advice of physicians. The reader should regularly consult a physician in matters relating to their health, and particularly with respect to any symptoms that may require diagnosis or medical attention.

All images used in this book are either royalty-free, in the public domain, or sourced from various websites with proper attribution and links back to the original owner websites where applicable. These images are used for educational and illustrative purposes only, with full respect to the rights of the original creators. The cover design and select illustrations were created by the talented creative team at AB Publisher LLC, an educational technology and publishing company. The written content was developed by the expert creators and educators at AB Publisher LLC, with the goal of providing accessible and high-quality educational material to learners worldwide. For any inquiry, please contact: ABPublisherLLC@Gmail.com

Table of Contents

0. Preface

0.1.What is Biology?

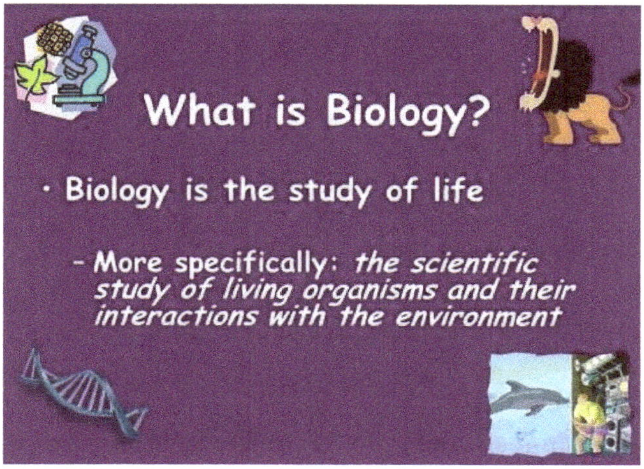

Figure 0.1: Definition of Biology

Biology is the scientific study of life — from the tiniest microbes to the largest ecosystems on Earth. It is a field that seeks to understand the structure, function, growth, origin, evolution, and interactions of living organisms. Biology helps us answer fundamental questions about life, from how cells work to how species evolve over time, and even how life itself began.

0.2. The Origin of the Term "Biology"

The word **biology** comes from the Greek words "bios" (life) and "logos" (study). This term, coined in the early 19th century, reflects humanity's deep and enduring curiosity about the natural world. Ancient scholars such as Aristotle made early observations about plants and animals, laying the groundwork for biology as a formal science.

0.3. Nature of Biology as a Discipline

Biology is a vast and interdisciplinary science, connecting with fields such as chemistry, physics, environmental science, medicine, and engineering. Its focus ranges from the molecular and cellular level — the microscopic machinery of life — to entire ecosystems, where species interact in complex webs of life.

0.4. Key Characteristics of Biology

1. **Diversity of Life Forms**
   Biology studies organisms of all shapes and sizes, from viruses and bacteria to plants, animals, and fungi.

2. **Adaptation and Evolution**
   Biology explains how species change over generations in response to environmental pressures — a concept central to the theory of evolution.

3. **Structure and Function**
   Biological structures, from cellular organelles to tissues and organs, are intimately linked to their functions.

4. **Energy and Metabolism**
   All life forms require energy to survive, grow, and reproduce — processes governed by biological and chemical laws.

5. **Homeostasis**
   Living organisms maintain stable internal environments despite changes in external conditions.

0.5. Divisions of Biology

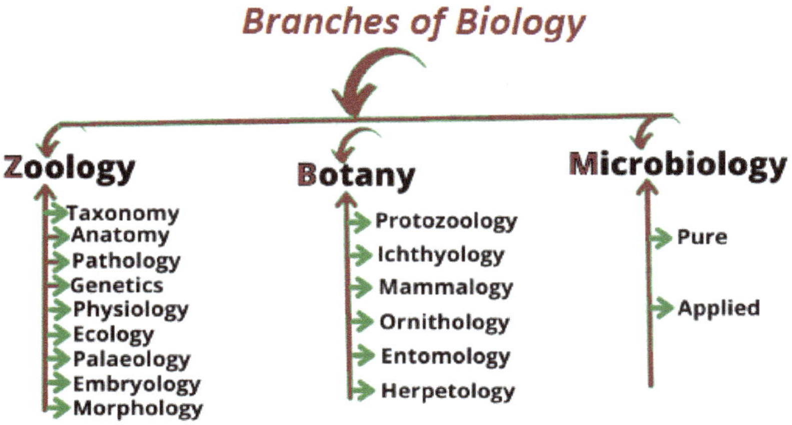

Figure 0.2: Some divisions of Biology

Biology is so broad that it is subdivided into specialized branches, each focusing on a particular aspect of life. Some key divisions include:

- **Cell Biology**: Study of cells, their structures, and functions.
- **Molecular Biology**: Study of biological molecules (like DNA, RNA, and proteins) and how they contribute to life processes.
- **Genetics**: Study of genes, heredity, and variation.
- **Ecology**: Study of organisms in relation to their environments.
- **Evolutionary Biology**: Study of how species change over time.
- **Microbiology**: Study of microorganisms like bacteria and viruses.
- **Botany**: Study of plants.
- **Zoology**: Study of animals.

0.6. Methods of Study in Biology

Biologists use a wide range of methods and tools to study life, including:

- **Observation and Experimentation**
  Direct observation of organisms in their natural environments and controlled experiments in laboratories help uncover biological processes.

- **Microscopy**
  Advanced microscopes reveal cellular structures and molecular details invisible to the naked eye.

- **Molecular Techniques**
  Tools like PCR, gene sequencing, and CRISPR allow scientists to study and even modify genes at the molecular level.

0.7. Scales of Study in Biology

The study of life spans several levels of biological organization:

- **Molecular Level**: DNA, RNA, proteins, and other biomolecules.
- **Cellular Level**: Individual cells, the fundamental units of life.
- **Tissues and Organs**: Groups of cells working together.
- **Organisms**: Complete living beings.
- **Populations**: Groups of the same species.
- **Communities and Ecosystems**: Interactions between species and their environments.
- **Biosphere**: All life on Earth.

0.8. Why Biology Matters

Biology is not just a subject in textbooks — it is central to understanding health, medicine, agriculture, environmental protection, biotechnology, and even space exploration. Whether solving global challenges like climate change or developing new medical treatments, biology holds the keys to improving life on Earth and beyond.

This preface serves as the foundation for the rest of the book, which will guide you through the fascinating world of cell and molecular biology — the study of life's inner workings at the smallest scales. From atoms and macromolecules to cell membranes, DNA replication, and gene expression, you’ll gain a deep understanding of the processes that make life possible.

1. Why Study Cell and Molecular Biology?

Life, in all its complexity and diversity, begins at the cellular and molecular levels. Every living organism is either a single cell or made up of trillions of cells working together. These cells, in turn, are built from intricate molecules — DNA, proteins, lipids, and more — that perform countless tasks essential for survival. To truly understand how life works, one must first understand how cells function and how molecules interact within those cells.

1.1. The Foundation of All Life

At its core, cell and molecular biology is the study of life's fundamental unit — the cell — and the molecular processes that drive its function. From bacteria to plants to humans, all life forms are made of cells, and all cells follow common molecular principles to sustain life. The diagram in Figure 1.1 is a clear generalized **prokaryotic** and **eukaryotic cell**, showing organelles and molecular components.

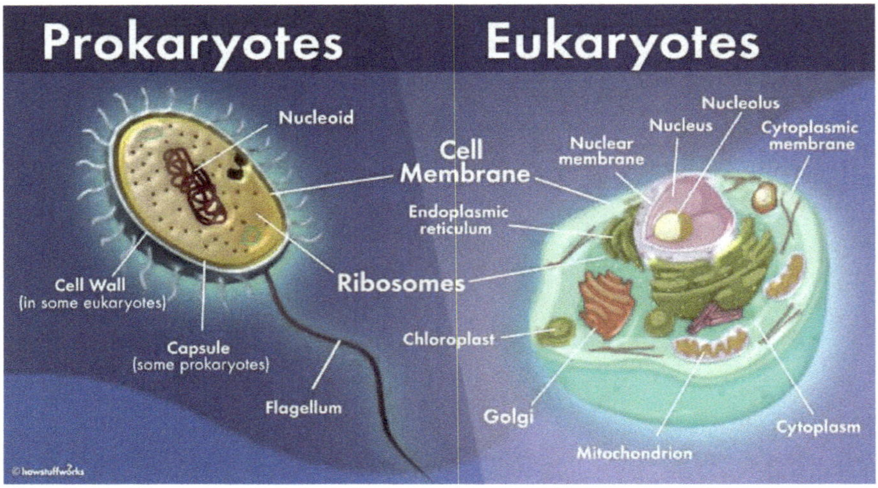

*Figure 1.1: A generalized **prokaryotic and eukaryotic cell**, showing organelles and molecular components: Source: https://science.howstuffworks.com/life/cellular-microscopic/prokaryotic-vs-eukaryotic-cells.htm*

Why is this important?

- **Health and Medicine**: Understanding cells helps us fight diseases, develop new drugs, and design advanced therapies like **gene editing** and **personalized medicine**.
- **Biotechnology and Innovation**: Molecular biology techniques allow us to engineer crops, develop vaccines, and produce biofuels.
- **Evolution and Origins**: By studying molecules within cells, we trace the evolutionary history of life itself.
- **Environmental Impact**: Microbial cells play key roles in nutrient cycling, waste decomposition, and climate regulation.

1.2. Connecting Biology, Chemistry, and Physics

Cell and molecular biology bridges the gap between biology and the physical sciences. Biological molecules follow the same **chemical and physical laws** that govern non-living matter, but they assemble into far more complex and purposeful systems. Understanding how biomolecules fold, interact, and react is essential to uncovering the secrets of life. The diagram in Figure 1.2 shows a molecular interaction, such as enzyme-substrate binding, with chemical bonds highlighted.

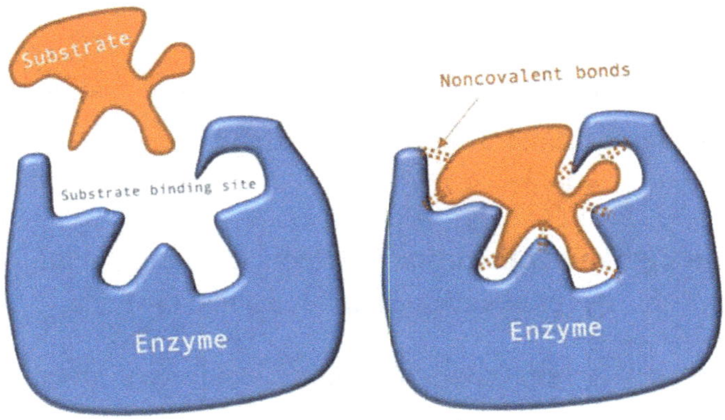

Figure 1.2: An *illustration of how enzymes bind to substrates — a perfect example of molecular interactions governed by physics and chemistry. Source: https://www.futurelearn.com/info/courses/the-biology-of-bugs-brains-and-beasts/0/steps/68820*

Examples of Molecular-Level Insights

- Enzyme activity depends on chemical bonds and reaction rates.
- Membrane transport relies on concentration gradients and diffusion principles.
- Gene expression depends on the physical interactions between DNA, RNA, and proteins.

1.3. Medical Applications: Understanding Disease at the Cellular Level

Every disease — from cancer to viral infections — involves disruptions at the cellular or molecular level. By studying how cells function normally and how they malfunction, scientists can:

- **Diagnose diseases more accurately** using genetic and molecular biomarkers.
- **Design targeted therapies** that correct or block harmful molecular processes.

- **Predict disease risk** based on individual genetic profiles.
- **Develop vaccines and antiviral drugs** by understanding how pathogens invade cells and hijack molecular machinery.

1.4. Revolutionary Tools and Techniques

Modern biology depends heavily on cell and molecular biology techniques. These tools allow scientists to observe, manipulate, and analyze cells and molecules with extraordinary precision. Some groundbreaking techniques include:

- **CRISPR-Cas9 Gene Editing**: Precisely altering DNA sequences. Figure 1.3 shows how the CRISPR-Cas9 system works — a modern tool in molecular biology.
- **Fluorescence Microscopy**: Watching molecules move inside living cells in real time.
- **Polymerase Chain Reaction (PCR)**: Amplifying tiny amounts of DNA for study.
- **Next-Generation Sequencing (NGS)**: Reading entire genomes quickly and affordably.
- **Single-Cell Analysis**: Studying individual cells to reveal cellular diversity.

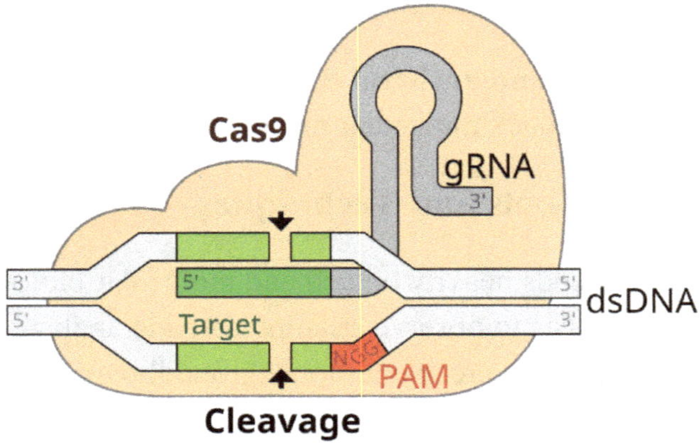

Figure 1.3: An illustration of ***CRISPR-Cas9 gene editing*** *process. Source: https://en.wikipedia.org/wiki/CRISPR_gene_editing*

These technologies are not just used in research labs — they are transforming medicine, agriculture, forensics, and environmental science.

1.5. Evolution and the Unity of Life

Cell and molecular biology also reveal the shared heritage of all living things. Whether you're studying a bacterium, a tree, or a human, you'll find the same types of molecules, the same genetic code, and many of the same cellular processes. This remarkable unity is strong evidence for the evolutionary connection between all life forms. Figure 1.4 shows a widely-used tree of life diagram perfect for illustrating evolutionary connections at the cellular level.

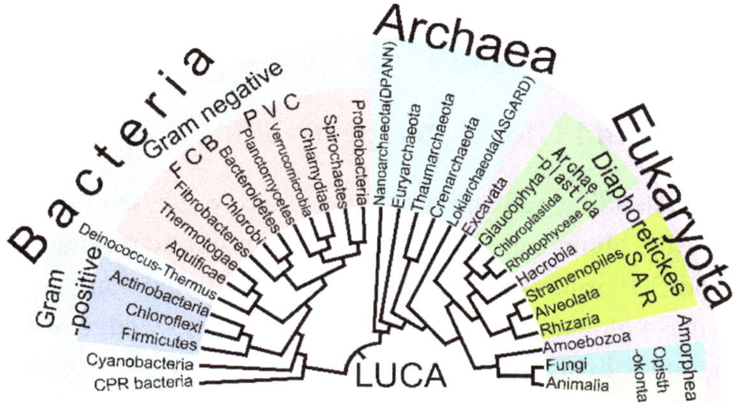

*Figure 1.4: A **phylogenetic tree of life**, showing the evolutionary relationship between species. Source: https://commons.wikimedia.org/wiki/File:Phylogenetic_Tree_of_Life.png*

Example

- The process of DNA replication works similarly in bacteria and humans.
- All cells use ATP as an energy currency.
- Proteins in photosynthesis and cellular respiration are found across species.

1.6. The Future of Science and Technology

The next wave of scientific breakthroughs will rely heavily on cell and molecular biology. Some areas where this knowledge will shape the future include:

- **Regenerative Medicine**: Using stem cells to repair or replace damaged tissues.
- **Synthetic Biology**: Designing entirely new biological systems for useful purposes.
- **Space Exploration**: Understanding how cells adapt to microgravity and cosmic radiation.

- **Personalized Genomics**: Tailoring medical treatments based on individual genetic profiles.

1.7. A Personal and Global Perspective

Finally, understanding cell and molecular biology is essential for informed citizens. Issues like climate change, pandemics, antibiotic resistance, and genetically modified organisms (GMOs) all have biological underpinnings. To participate in these global conversations, a working knowledge of how cells and molecules operate is invaluable.

Key Takeaway

As you embark on this journey through cell and molecular biology, remember that you are exploring the blueprints of life itself. Every breath you take, every meal you eat, and every thought you have depends on the microscopic, molecular processes happening within your own cells — and this book will help you unlock those secrets.

2. The Study of Evolution

2.1. Theory of Evolution

Evolution is the process by which populations of organisms change over generations through genetic variation and natural selection. The **Theory of Evolution**, first formalized by **Charles Darwin** in his 1859 book *On the Origin of Species*, is one of the most foundational concepts in biology. It explains how all life on Earth shares common ancestry and how species diverge over time to adapt to their environments.

At its core, the **theory rests on two key principles**:

- **Descent with modification:** Populations change over generations, and offspring inherit traits from their parents, though mutations and recombination introduce variation.
- **Natural selection:** Individuals with advantageous traits are more likely to survive and reproduce, passing those beneficial traits to the next generation.

Modern evolutionary theory has incorporated knowledge from genetics, molecular biology, and population biology, making it a cornerstone in understanding biodiversity and the origin of species.

The Tree of Life diagram (showing evolutionary relationships across species) can be seen here**:**
https://commons.wikimedia.org/wiki/File:Tree_of_life_SVG.svg

2.2. Adaptive Evolution

Adaptive evolution refers to changes in populations driven by **natural selection**, where organisms better suited to their environment are more likely to survive and reproduce. This adaptation often occurs in response to:

- **Environmental pressures**, such as changes in climate, availability of resources, or competition.

- **Predation and disease**, where adaptations may involve camouflage, resistance to pathogens, or changes in reproductive strategies.

Example:

The development of antibiotic resistance in bacteria is a real-time example of adaptive evolution. Bacteria with genes that provide resistance to antibiotics survive and pass these traits to future generations.

2.3. Convergent Evolution

Convergent evolution occurs when unrelated species evolve similar traits independently, usually because they occupy similar ecological niches or face comparable environmental challenges.

Examples:

- The **wings of bats and birds** evolved independently to serve the same purpose — flight.
- **Desert plants in Africa and North America**, like cacti and euphorbias, evolved similar water-storing adaptations despite being unrelated.

Figure 2.1 shows an example of convergent evolution (placental vs marsupial mammals).

Figure 2.1: Example of convergent evolution (placental vs marsupial mammals). Source: https://www.daviddarling.info/encyclopedia/C/convergevol.html

2.4. Natural Selection

Natural selection is the process by which certain heritable traits become more common in a population because they enhance an organism's survival and reproductive success. It is the primary mechanism of evolution.

Key Elements of Natural Selection:

- **Variation**: Individuals within a species show differences.
- **Heritability**: Traits can be passed from parents to offspring.
- **Differential Survival and Reproduction**: Individuals with advantageous traits survive longer and reproduce more.
- **Adaptation**: Over time, the population shifts to reflect these advantageous traits.

Example:

The **peppered moth** during the Industrial Revolution in England is a classic case. Dark-colored moths became more common in polluted areas because they were better camouflaged against soot-darkened tree trunks, making them less visible to predators.

Figure 2.2: Peppered moth (example of natural selection). Source: https://butterfly-conservation.org/moths/why-moths-matter/amazing-moths/peppered-moth-and-natural-selection

2.5. Vestiges of Evolution

Vestigial structures are anatomical features or behaviors that no longer serve their original purpose, but are remnants of evolutionary history. These structures offer evidence of common ancestry.

Examples:

- **Human tailbone (coccyx)**: A remnant of tails found in our primate ancestors.
- **Whale pelvis bones**: Remnants from ancestors that walked on land.

- **Goosebumps in humans**: A leftover response from ancestors with fur, where this reflex helped puff up their coats for warmth or intimidation.

2.6. Methods of Studying Evolution

Modern biologists use multiple scientific approaches to study evolutionary processes, combining fossil records, genetics, and observational studies.

Key Methods:

1. **Fossil Evidence**
   Fossils provide a record of past life forms and how species have changed over time.

2. **Comparative Anatomy**
   Studying anatomical similarities and differences helps reveal evolutionary relationships.

3. **Genetic Analysis**
   Advances in molecular biology allow scientists to compare DNA sequences across species, uncovering genetic similarities that confirm evolutionary relationships.

4. **Observational Studies**
   Real-time studies of evolving populations, such as Darwin's finches in the Galápagos Islands, demonstrate evolution in action.

5. **Experimental Evolution**
   Scientists can observe evolution directly in controlled environments, such as bacterial populations exposed to antibiotics or fruit flies selected for specific traits.

6. **Biogeography**
   The geographic distribution of species helps scientists infer evolutionary processes. Example: Unique species on islands, like the marsupials of Australia, evolved in isolation, providing clues about migration and

speciation.

3. What is Cell Biology?

3.1. Introduction to Cell Biology

Cell biology is the branch of biology focused on the structure, function, and behavior of cells — the fundamental units of life. All living things, from bacteria to humans, are composed of cells. Understanding cells allows scientists to explore the mechanisms underlying life processes, from gene expression to energy production.

Cell biology is a foundational discipline within biology because every biological process — including growth, development, reproduction, and metabolism — depends on the activities of cells.

Key Focus Areas of Cell Biology:

- **Cell structure** — the organization of organelles, membranes, and cytoskeleton.
- **Cell function** — processes like nutrient uptake, energy production, and waste removal.
- **Cell communication** — how cells send and receive signals from their environment.
- **Cell division and reproduction** — how cells multiply and maintain genetic information.
- **Cell evolution** — how cells have evolved from simpler forms into complex life forms.

3.2. Importance of Cell Biology

Studying cells helps researchers understand the origins of life, how diseases develop, and how organisms function at the most basic level. Modern medicine, biotechnology, and even environmental science heavily rely on advances in cell biology.

Examples of Cell Biology Applications:

- **Cancer research** — Understanding how cell division becomes unregulated.
- **Genetic engineering** — Modifying cellular DNA for agriculture or medicine.
- **Drug development** — Targeting cellular processes to design new treatments.
- **Regenerative medicine** — Using stem cells to repair or replace damaged tissues.

3.3. Historical Background

The discovery of cells dates back to 1665, when Robert Hooke used a primitive microscope to observe thin slices of cork and described the "tiny boxes" he saw as "cells." Since then, technological advancements, from light microscopes to electron microscopes and live-cell imaging, have provided increasingly detailed views of cell structures and processes.

The Cell Theory

The Cell Theory, a cornerstone of biology, consists of three main tenets:

1. All living organisms are composed of one or more cells.
2. The cell is the basic unit of structure and function in all living things.
3. All cells arise from pre-existing cells.

3.4. Types of Cells: Prokaryotic vs. Eukaryotic

Cells fall into two broad categories:

- **Prokaryotic Cells:** Simpler, lacking a nucleus or membrane-bound organelles (e.g., bacteria and archaea).
- **Eukaryotic Cells:** More complex, with a **nucleus** and **organelles** (e.g., plants, animals, fungi, and protists).

Table 3.1. Comparison of Prokaryotic and Eukaryotic Cells

Feature	Prokaryotic Cells	Eukaryotic Cells
Nucleus	Absent	Present
Size	Smaller (0.1-5 μm)	Larger (10-100 μm)
Organelles	Few	Numerous
DNA	Circular	Linear
Cell Division	Binary Fission	Mitosis/Meiosis
Examples	Bacteria, Archaea	Plants, Animals, Fungi, Protists

3.5. Multicellular Organisms

Most of the life we observe — from plants and animals to fungi — is multicellular, meaning their bodies consist of many specialized cells working together. Each cell in a multicellular organism plays a specific role, contributing to the overall function and survival of the organism.

Characteristics of Multicellular Organisms

- **Cell Specialization (Differentiation):** Cells develop specific functions — e.g., muscle cells contract, nerve cells transmit signals.
- **Tissue Formation:** Specialized cells group together to form tissues (e.g., muscle tissue, nervous tissue).
- **Organ Systems:** Tissues combine to form organs, and organs work together in organ systems (e.g., digestive system, circulatory system).
- **Cell Communication:** Cells in multicellular organisms must constantly **communicate** to coordinate processes like growth, immune responses, and homeostasis.

Example: Human Body

- **Muscle Cells** contract to enable movement.
- **Red Blood Cells** transport oxygen.
- **Neurons** transmit signals.
- **Epithelial Cells** form protective layers.
- Together, these specialized cells allow the human body to function as a **coordinated whole**.

3.6. Evolution of Multicellularity

Multicellularity evolved multiple times in Earth's history, allowing the development of complex life forms. Cells gained the ability to adhere to each other, communicate, and differentiate, which led to the evolution of large, complex organisms.

Key Steps in the Evolution of Multicellular Life:

1. **Aggregation of Cells** — Single cells grouped together.
2. **Division of Labor** — Some cells specialized for particular functions.
3. **Genetic Control of Development** — Genes evolved to guide the formation of tissues and organs.

Chapter Summary

- Cell biology focuses on understanding the structure and function of cells.
- Cells are either **prokaryotic** or **eukaryotic**, with distinct characteristics.
- Multicellular organisms are composed of specialized cells working together.
- Studying cell biology is fundamental to understanding life, health, disease, and biotechnology.

4. Genetics and Our Genetic Blueprints

4.1. Introduction

At the heart of all life processes lies genetics, the study of how traits are inherited and how information is encoded within organisms. Every cell in your body, every characteristic you display — from eye color to susceptibility to diseases — is rooted in the genetic blueprint stored within DNA.

The field of genetics explains how traits are passed across generations, how cells function and communicate, and even how species evolve over time.

4.2. Evolution and Genetics

Evolution is driven by genetic changes that accumulate across generations. Mutations, natural selection, and genetic drift act upon the genetic material, leading to adaptations and diversification.

- **Mutation:** Random changes in DNA sequences, which can lead to new traits.
- **Natural Selection:** Genetic variations that improve survival and reproduction become more common in populations.
- **Speciation:** Genetic divergence between populations eventually results in new species.

4.3. Developmental Encoding

The **genetic code** doesn't just control individual traits; it also encodes how an organism develops from a single fertilized egg into a complex body. This process, called developmental encoding, relies on:

- **Gene Regulation:** Different genes are turned on or off at different stages.
- **Signaling Pathways:** Cells communicate to coordinate development.

- **Morphogenesis:** Genes guide the formation of tissues and organs.

4.4. DNA – The Molecular Blueprint

DNA, or **deoxyribonucleic acid**, is the molecule that carries genetic information. Its famous **double helix** structure holds the code for building proteins, which in turn, carry out the functions of life.

Key Features of DNA:

- **Nucleotide Bases:** Adenine (A), Thymine (T), Cytosine (C), Guanine (G).
- **Base Pairing:** A pairs with T, C pairs with G.
- **Genes:** Sections of DNA that encode proteins.
- **Chromosomes:** DNA is organized into structures called chromosomes.

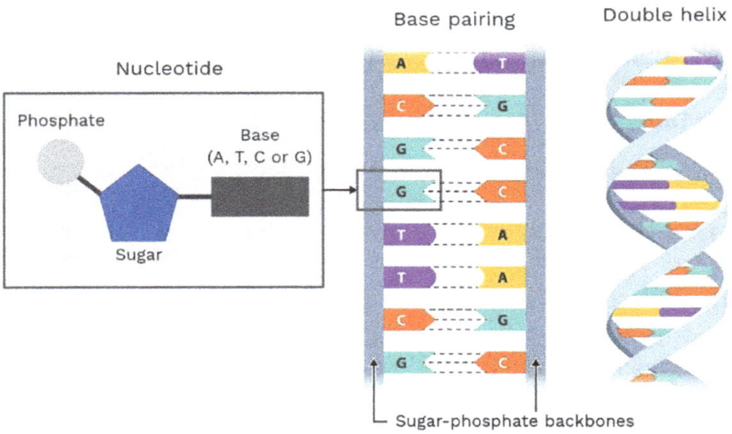

Figure 4.1: Diagram of DNA structure with nucleotide bases labeled. Source: https://theory.labster.com/structure-dna/

4.5. LINE-1 Elements – Jumping Genes

Not all parts of the genome directly code for proteins. Some regions, such as **LINE-1 elements**, are **transposable elements** — sequences of DNA that can copy and insert themselves into new locations within the genome.

Why Are LINE-1 Elements Important?

- **Genome Evolution:** They contribute to **genome diversity**.
- **Gene Regulation:** Their movement can **activate or silence nearby genes**.
- **Disease:** LINE-1 activity has been linked to some **cancers** and **neurological disorders**.

4.6 Epigenetics – Beyond the Genetic Code

Epigenetics refers to heritable changes in gene activity that occur without altering the DNA sequence itself. These changes are influenced by **environment, lifestyle, and development**.

Common Epigenetic Mechanisms:

- **DNA Methylation:** Adding methyl groups to DNA can **silence genes**.
- **Histone Modification:** Proteins around which DNA is wrapped are chemically modified to turn genes on or off.
- **Non-coding RNA:** Some RNA molecules help regulate gene expression.

Epigenetics and Disease

Epigenetic changes have been linked to:

- **Cancer** — Silencing tumor suppressor genes.
- **Neurological disorders** — Abnormal gene regulation.
- **Aging** — Changes in gene expression over time.

4.7. Disease-gene Associations

The study of disease-gene associations helps scientists understand how genetic variations contribute to disease. Advances in genome-wide association studies (GWAS) have identified thousands of genes linked to conditions like:

- Diabetes
- Heart disease
- Cancer
- Neurological disorders

Genetic Risk vs. Environmental Influence

Some diseases are **purely genetic**, caused by mutations in single genes (like **cystic fibrosis**). Others arise from a **complex interplay** of genetic predisposition and environmental factors.

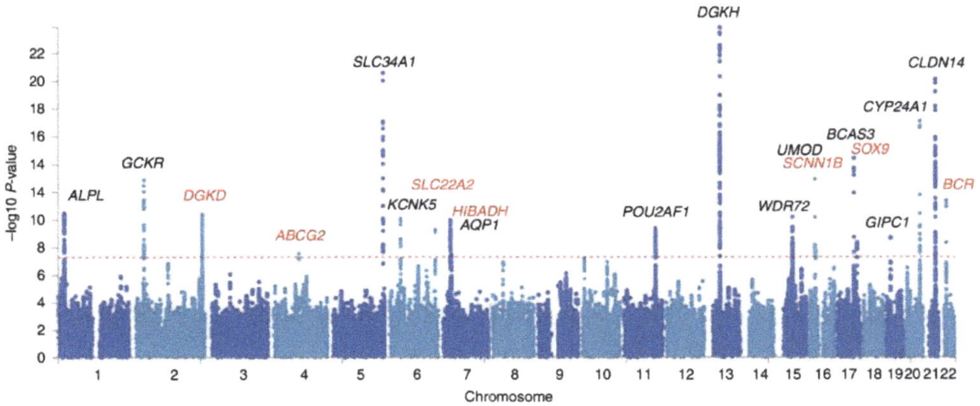

Figure 4.2: Manhattan plot from a GWAS of kidney stone disease. Source: https://en.wikipedia.org/wiki/Genome-wide_association_study

4.8. Reverse Engineering the Genome

Scientists increasingly reverse engineer genetic information to understand evolution, disease, and development. This involves:

- **Comparative Genomics:** Comparing genomes across species to find **conserved genes**.
- **Gene Editing:** Techniques like **CRISPR** allow scientists to edit genes directly, unlocking the potential for gene therapy.
- **Synthetic Biology:** Scientists can design new genes or even entire synthetic organisms.

Here is the link to an illustration of CRISPR gene editing process: https://www.genome.gov/genetics-glossary/CRISPR

Chapter Summary

- Genetics is the **foundation of biology**, governing inheritance, development, and evolution.
- **DNA** encodes the instructions for life, while **epigenetics** controls how those instructions are used.
- **Transposable elements** like LINE-1 shape genomes through their movement.
- Modern tools like **GWAS** and **CRISPR** help us understand **disease-gene associations** and develop **new therapies**.
- Understanding the genetic blueprint is key to understanding **life itself**.

5. Getting Down with Atoms

5.1. Introduction

At the most fundamental level, all life — from the simplest bacterium to the largest whale — is built from **atoms**. Understanding atoms is essential to understanding **biology**, since atoms combine to form **molecules**, which form **cells**, and ultimately, entire organisms.

Atoms are the basic units of matter. They consist of a **central nucleus**, which contains protons and neutrons, and **electrons** that orbit around this nucleus.

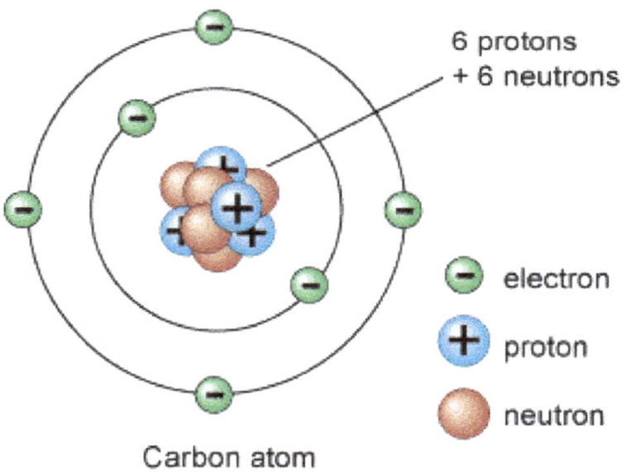

*Figure 5.1: An atom consists of a **central nucleus**, which contains protons and neutrons, and **electrons** that orbit around this nucleus. Source: https://www.universetoday.com/56469/atom-diagram/*

The properties of atoms determine how they interact to form the molecules of life, including DNA, proteins, and carbohydrates.

5.2. Subatomic Particles

Every atom is made of three primary subatomic particles, each with distinct properties.

Table 5.1: Properties of Subatomic Particles

Particle	Charge	Relative Mass	Location
Proton	Positive (+)	1	Nucleus
Neutron	Neutral (0)	1	Nucleus
Electron	Negative (-)	Approximately 0	Orbiting the nucleus

These particles define the atom's identity and behavior. For example, the number of protons in the nucleus determines the element.

5.3. The Atomic Nucleus

At the core of every atom lies the **atomic nucleus**, a compact region that contains **protons** and **neutrons**.

Protons

- Positive charge (+1).
- Determine atomic number — the number of protons defines what element an atom is (e.g., carbon has 6 protons).

Neutrons

- No charge (neutral).
- Stabilize the nucleus by spacing out the positively charged protons, which would otherwise repel each other.
- Varying numbers of neutrons create **isotopes** — atoms of the same element with different masses.

The nucleus is incredibly dense — nearly all the atom's mass resides here.

5.4. Electrons – The Movers and Shakers

Electrons are tiny, negatively charged particles that orbit the nucleus in regions called electron shells or orbitals. They are crucial to chemical bonding and biological interactions.

Key Facts About Electrons

- **Mass:** Almost negligible compared to protons and neutrons.
- **Location:** Move within probabilistic regions called orbitals, not fixed paths.
- **Charge:** Negative (-1), balancing the positive charge of protons.
- **Chemical Reactions:** Electrons are responsible for forming chemical bonds, which allow atoms to combine into molecules.

Valence Electrons

The electrons in the outermost shell (called **valence electrons**) are the ones that participate in chemical reactions. Atoms gain, lose, or share these electrons to achieve stability, forming:

- **Ionic bonds** (transfer of electrons).
- **Covalent bonds** (sharing of electrons).

5.5. Neutrons – The Silent Stabilizers

While protons define the element, neutrons help hold the nucleus together. Without neutrons, the repulsive forces between protons would blow the nucleus apart. The balance between protons and neutrons determines nuclear stability.

Isotopes

Isotopes are atoms of the same element with different numbers of neutrons. For example, carbon-12 and carbon-14 are both carbon isotopes, but carbon-14 has two extra neutrons. Some isotopes are stable, while others are radioactive and decay over time.

Chapter Summary

- **Atoms** are the building blocks of all biological molecules.
- Each atom consists of **protons**, **neutrons**, and **electrons**, each playing a distinct role.
- **The nucleus** holds protons (positively charged) and neutrons (neutral), giving the atom its **mass**.
- **Electrons** (negatively charged) orbit the nucleus and drive **chemical reactions**.
- **Neutrons** provide nuclear stability and create **isotopes**, some of which are used in biological research and medicine.

6. How Chemical Bonds Combine Atoms

The molecules that make up all living things — including you — are held together by **chemical bonds**. These bonds form between atoms to create stability and allow complex structures to arise from simple elements. Understanding these bonds is essential to understanding how biological molecules function.

6.1. Covalent Bond

A **covalent bond** forms when **two atoms share one or more pairs of electrons**. This sharing allows both atoms to achieve a more stable electronic configuration, often filling their outer electron shells. Covalent bonds are particularly important in biological molecules such as **DNA, proteins, and carbohydrates**, where **carbon atoms** bond covalently with hydrogen, oxygen, nitrogen, and other atoms.

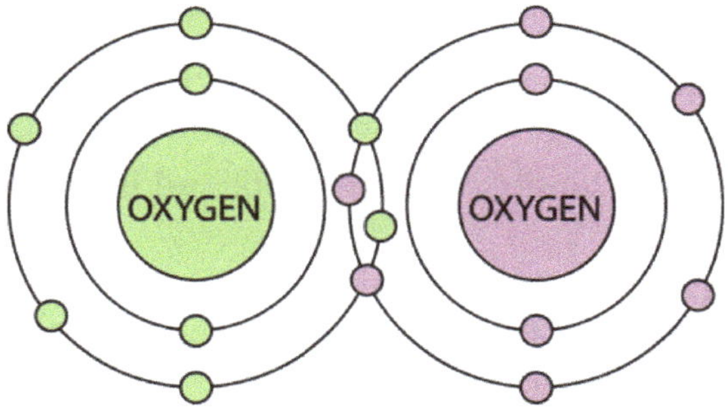

Figure 6.1: ***Two oxygen atoms sharing electrons to form a covalent bond.***

6.2. Ionic Bond

An **ionic bond** forms when one atom **donates** one or more electrons to another atom, creating **ions** — charged atoms. The atom that loses electrons becomes a **positively charged ion (cation)**, while the atom that gains electrons becomes a **negatively charged ion (anion)**. The **electrostatic attraction** between these oppositely charged ions forms the ionic bond.

For example, **sodium (Na)** donates one electron to **chlorine (Cl)**, forming **Na^+** and **Cl^-**, which combine to form **sodium chloride (NaCl)** — common table salt.

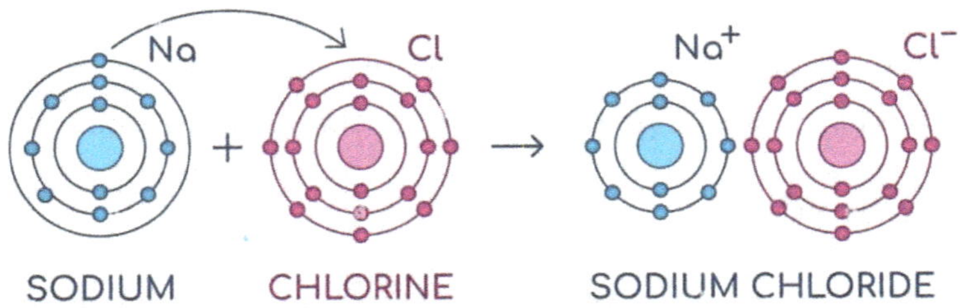

Figure 6.2: Illustration showing electron transfer and formation of ionic bond in sodium chloride.

6.3. Metallic Bond

A **metallic bond** occurs in metals, where atoms share a **"sea of electrons"** that flows freely around and between atoms. This free movement of electrons gives metals their unique properties, such as **electrical conductivity**, **malleability**, and **luster**. While metallic bonds are less relevant to biological molecules, they are important in understanding **biomedical implants and instruments** made from metals.

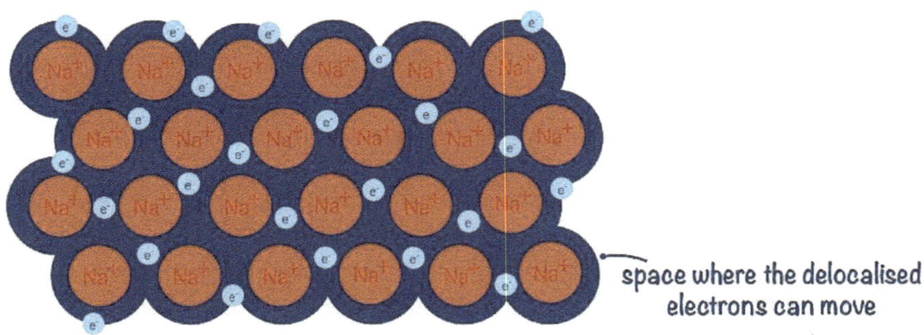

Figure 6.3: Illustration of metallic bonding with free-moving electrons. Source: https://www.chemistrystudent.com/metallicbonding.html

6.4. Single Bond

A **single bond** is a type of covalent bond where **only one pair of electrons** is shared between two atoms. Single bonds are common in organic molecules, where carbon atoms form **single bonds** with hydrogen atoms and other carbon atoms.

Example: **H—H (Hydrogen molecule)**.

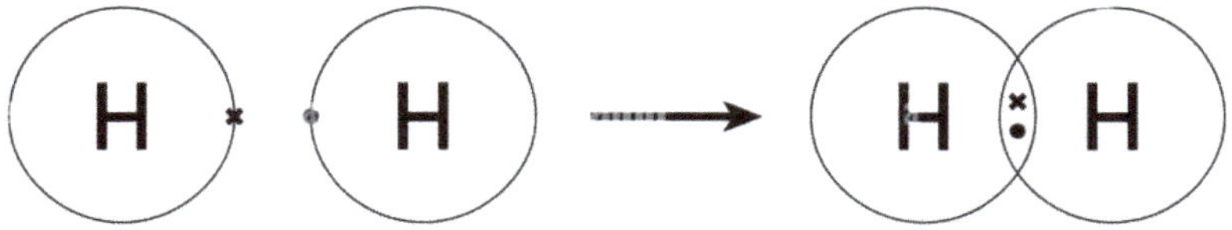

Figure 6.4: Illustration of a single bond between hydrogen atoms.

6.5. Double Bond

A **double bond** forms when **two pairs of electrons** are shared between two atoms. Double bonds are **stronger** than single bonds and influence the **shape and properties** of molecules.

Example: **O=O (Oxygen molecule)**.

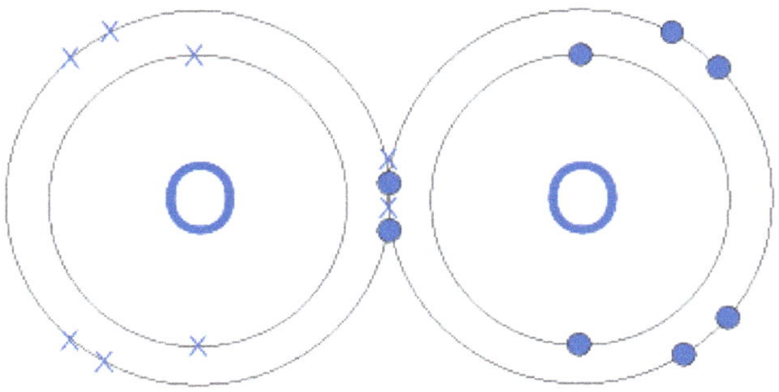

Figure 6.5: Illustration of a double bond between oxygen atoms.

6.6. Electrostatic Force

Electrostatic forces are the **forces of attraction or repulsion between charged particles**. These forces play a critical role in **ionic bonding** and influence how charged molecules interact inside biological systems.

6.7. Electrostatic Attraction

This is the **force that pulls opposite charges toward each other**. Electrostatic attraction is essential in ionic bonds and also helps stabilize protein structures, where oppositely charged side chains of amino acids attract each other.

6.8. Electrostatic Repulsive Force

The electrostatic repulsive force occurs when particles with the same charge push each other apart. This is crucial in biological systems, such as the phosphate groups in DNA, which are negatively charged and repel each other — a factor influencing the helical structure of the DNA molecule.

7. Water, Solutions and Mixtures

Life as we know it depends heavily on water and the various solutions and mixtures formed with it. Water is sometimes called the **universal solvent** because it dissolves more substances than any other liquid. In living organisms, biological processes rely heavily on the formation of solutions, where substances are dissolved and transported within cells and tissues. In this chapter, we'll explore water's properties, the concept of **solutions**, and the nature of **mixtures**.

7.1. Intermolecular Forces

Intermolecular forces are the **forces of attraction or repulsion between molecules**. These forces are weaker than covalent or ionic bonds, but they still play a critical role in how substances dissolve and how mixtures behave.

There are three primary types of intermolecular forces:

- **Hydrogen bonding** – Attraction between a hydrogen atom covalently bonded to a highly electronegative atom (like oxygen or nitrogen) and another electronegative atom nearby. This is particularly important in water.
- **Dipole-dipole interactions** – Attractions between polar molecules, where partial positive and partial negative charges attract.
- **Van der Waals forces** – Weak **attractions between nonpolar molecules** caused by **temporary dipoles**.

7.2 Solubility of a Solute

Solubility refers to how much of a **solute** (the substance being dissolved) can dissolve in a given amount of **solvent** (the dissolving medium), often water. Solubility is influenced by:

- **Temperature** – In most cases, increasing temperature increases solubility of solids but decreases solubility of gases.

- **Polarity** – The **"like dissolves like"** rule states that **polar solutes** dissolve well in **polar solvents** (like water), while **nonpolar solutes** dissolve better in **nonpolar solvents** (like oil).

7.3. Ionic Nature of Liquid Solutions

Many biological solutions, like **blood plasma** and **cytoplasm**, are **electrolyte solutions**, meaning they contain ions that conduct electricity. When an **ionic compound** dissolves in water, it dissociates into positive and negative ions, which are surrounded by water molecules in a process called **hydration**.

For example, dissolving **NaCl** (table salt) in water results in:

$$NaCl \rightarrow Na^{+} + Cl^{-}$$

The ability to form **ionic solutions** is crucial for nerve impulses, muscle contraction, and **pH regulation** in biological systems.

7.4. Stabilizing Agents in Liquid Solutions

In some mixtures, **stabilizing agents** help prevent separation or clumping of particles. These agents, also called **emulsifiers** or **surfactants**, have both polar and nonpolar regions, allowing them to bridge water-soluble and fat-soluble components.

Example: Bile salts act as emulsifiers in the digestive system, helping to break down fats into micelles that can be absorbed by the intestine.

7.5. Classifications of Mixtures

Mixtures are combinations of two or more substances that are **not chemically bonded**. They can be classified into:

- **Homogeneous mixtures (solutions)** – The composition is **uniform** throughout (e.g., saltwater).

- **Heterogeneous mixtures** – The composition is **not uniform**, and you can **see different components** (e.g., **oil and water**).
- **Colloids** – Particles are larger than in solutions but too small to settle out (e.g., **milk**).
- **Suspensions** – Large particles that eventually **settle out** (e.g., muddy water).

7.6. Chemical Misconceptions

When studying solutions and mixtures, common **misconceptions** include:

- Thinking dissolving means disappearance – In reality, dissolved substances are still present, just in molecular or ionic form.
- Confusing **melting** with **dissolving** – Melting is a **physical phase change**, while dissolving involves **interaction with a solvent**.
- Believing **more stirring always increases solubility** – Stirring **increases rate**, but not the **maximum amount** that can dissolve.

Correcting these misconceptions is important for a clear understanding of biological processes, where mixtures and solutions play essential roles.

8. Which Elements Are in Cells?

Cells, the basic units of life, are primarily composed of a handful of **essential elements** that form the building blocks of biological molecules. These elements combine in different ways to create proteins, lipids, carbohydrates, and nucleic acids, all of which are necessary for life. Let's explore some of the most important elements found in cells and their roles.

8.1. Carbon: The Backbone of Life

Carbon is often called the **element of life** because of its unique ability to form **four covalent bonds**, allowing it to create **complex chains, rings, and branches**. These structures serve as the backbone for organic molecules like proteins, lipids, and DNA.

In cells, **carbon** atoms bond with hydrogen, oxygen, nitrogen, and other elements to form **macromolecules** essential for life.

8.2. Hydrogen: Vital for Water and Organic Compounds

Hydrogen is the **most abundant element** in the universe and plays a crucial role in cells. It's part of water (H_2O), the medium where all biological processes occur. Hydrogen atoms also appear in nearly all biomolecules, including amino acids, fats, and nucleotides.

In addition, **hydrogen bonding** — weak interactions between hydrogen atoms and electronegative atoms like oxygen or nitrogen — helps stabilize the structure of DNA and proteins.

8.3. Hemoglobin: Iron and Oxygen Transport

While iron is not one of the most abundant elements in cells, it is **vital** for **hemoglobin**, the protein in red blood cells that carries oxygen throughout the body. Each hemoglobin molecule contains iron atoms, which bind to oxygen molecules, allowing cells to receive the oxygen they need for **cellular**

respiration.

8.4 Phosphorus: Energy and DNA Backbone

Phosphorus is essential for two major reasons:

1. **DNA and RNA Backbone** – Phosphate groups link nucleotides together to form the **sugar-phosphate backbone** of DNA and RNA.
2. **Energy Currency** – **Adenosine triphosphate (ATP)**, the cell's main energy carrier, contains three phosphate groups, which release energy when broken.

Without phosphorus, cells would lack the ability to store and transfer energy or encode genetic information.

8.5. Nitrogen: Key to Proteins and Nucleic Acids

Nitrogen is a critical component of amino acids, the **building blocks of proteins**, and **nucleotides**, the building blocks of DNA and RNA. Nitrogen's presence in these molecules allows cells to build enzymes, structural proteins, and genetic material.

In nature, **nitrogen fixation** — converting atmospheric nitrogen into forms usable by cells — is carried out by special bacteria that live in soil and plant roots.

These five elements — **carbon, hydrogen, phosphorus, nitrogen, and iron** — are only a few of the elements necessary for life, but they are among the most important. Together, they help **cells build molecules**, **store energy**, and **carry out essential processes**.

9. Macromolecules Are the "Big" Molecules in Living Things

Life depends on macromolecules, which are large, complex molecules built from smaller building blocks. These macromolecules carry out essential functions within cells — from storing genetic information to providing energy and building structural components.

9.1. Biomolecules: The Four Major Types

All living organisms rely on four main categories of **biomolecules**, which are also called **macromolecules**:

- **Carbohydrates**
- **Proteins**
- **Lipids**
- **Nucleic Acids**

These large molecules form from smaller **monomers**, which **bond** together through **polymerization**.

9.2. Functions of Macromolecules

Each type of **macromolecule** serves distinct functions critical to **cellular processes**:

- **Carbohydrates**: Provide energy and structural support.
- **Proteins**: Serve as enzymes, structural components, hormones, and more.
- **Lipids**: Store energy, form cell membranes, and provide insulation.
- **Nucleic Acids**: Store and transmit genetic information (DNA and RNA).

9.3. General Structure of Macromolecules

Though each type of macromolecule has a different shape and function, they share a common feature — they are **polymers**, meaning they are made up of repeating **monomer units**.

- **Carbohydrates** consist of **sugar (monosaccharide) units**.
- **Proteins** consist of **amino acids**.
- **Nucleic acids** consist of **nucleotides**.
- **Lipids** are **not true polymers**, but they often contain **long hydrocarbon chains** or **fatty acids**.

9.4. Key Elements in Macromolecules

Different **macromolecules** contain different combinations of essential **elements**:

- **Carbon (C)** — Forms the backbone of all **organic molecules**.
- **Hydrogen (H)** — Present in all macromolecules.
- **Oxygen (O)** — Especially important in **carbohydrates** and **nucleic acids**.
- **Nitrogen (N)** — Found in **proteins** (amino acids) and **nucleic acids**.
- **Phosphorus (P)** — A key component of **nucleic acids** and **cell membranes**.
- **Sulfur (S)** — Found in certain **amino acids** (like **cysteine** and **methionine**).

9.5. Carbohydrates: Quick Energy and Structural Support

Carbohydrates are macromolecules composed mainly of carbon, hydrogen, and oxygen, typically in a **1:2:1** ratio.

Main Functions

- **Energy Storage**: **Glucose**, a **simple sugar**, is a primary energy source for **cells**.
- **Structural Support**: **Cellulose** in **plants** and **chitin** in **fungi** and **insects** provide rigid structural support.

Types of Carbohydrates

- **Monosaccharides**: Simple sugars (e.g., **glucose**, **fructose**).
- **Disaccharides**: Two monosaccharides linked (e.g., **sucrose**).
- **Polysaccharides**: Long chains of monosaccharides (e.g., **starch**, **glycogen**, **cellulose**).

9.6. Nucleic Acids: Carriers of Genetic Instructions

Nucleic acids store and transmit genetic information in all living organisms. There are two main types:

- **Deoxyribonucleic Acid (DNA)** — Stores the genetic code.
- **Ribonucleic Acid (RNA)** — Helps convert the code into **proteins**.

Structure

Nucleic acids are made of nucleotides, each containing:

- A **phosphate group**
- A **sugar (deoxyribose or ribose)**
- A **nitrogenous base** (A, T, C, G, or U)

Functions

- **DNA** stores the instructions for making **proteins**.

- **RNA** helps **translate** these instructions into actual **proteins**.

Macromolecules are the foundation of **biological life**, enabling **cells** to store information, build structures, break down nutrients, and carry out essential **chemical reactions**. Understanding their **composition**, **structure**, and **function** is crucial to understanding how **life itself operates**.

10. Thermodynamics in Living Things

All **biological processes** — from photosynthesis to muscle contraction — rely on the flow and transformation of energy. Understanding how energy behaves in living systems requires knowledge of thermodynamics, the study of energy transfer and conversion.

10.1. The First and Second Laws of Thermodynamics in Living Things

Living organisms obey the **laws of thermodynamics**, just like all other systems in the **universe**. However, they exist in an **open system**, constantly exchanging matter and energy with their environment.

- **The First Law of Thermodynamics** (Law of Energy Conservation) states that **energy** cannot be created or destroyed, only transformed from one form to another. Example: In photosynthesis, light energy from the sun is converted into chemical energy stored in glucose.
- **The Second Law of Thermodynamics** states that energy transformations always increase the overall **entropy** (disorder) of the universe. This means some **energy** is always **lost** as **heat** during any **biological process**.

10.2. Gibbs Energy Change

To understand whether a reaction can occur spontaneously in cells, scientists use **Gibbs Free Energy Change (ΔG)**.

- If **ΔG** is **negative (<0)**, the process is **spontaneous** and can happen without additional **energy input**. This is called an **Exergonic** reaction.
- If **ΔG** is **positive (> 0)**, the process requires **energy input** to proceed. This is called an **Endergonic** reaction.
- Biological systems often couple reactions — using energy-releasing

reactions (like **ATP hydrolysis**) to drive energy-requiring reactions.

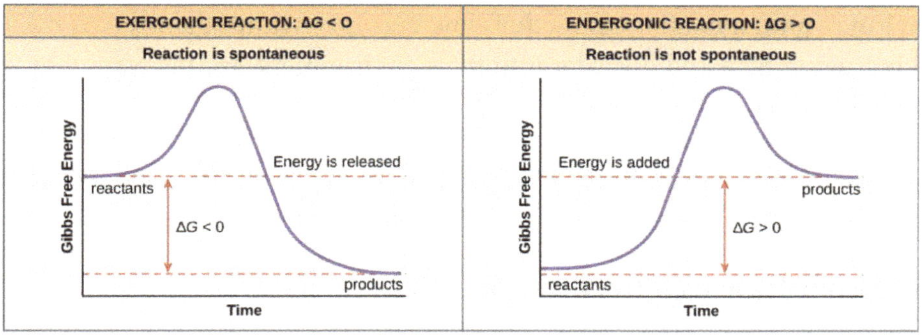

Figure 10.1: Illustration of the difference between Exergonic reaction ($\Delta G<0$) and Endergonic reaction ($\Delta G>0$). Source: https://www.khanacademy.org/science/biology/energy-and-enzymes/free-energy-tutorial/a/gibbs-free-energy

10.3. First Law of Thermodynamics

The First Law of Thermodynamics is a statement of **energy conservation**:

Energy in a **closed system** is constant. It can only **change form**, but the total amount remains the same.

In **living organisms**, **energy** flows continuously:

- **Light energy** enters through photosynthesis.
- **Chemical energy** (like glucose) powers metabolism.
- **Heat energy** leaves the system.

This constant input and output allow life to function despite the constraints of the First Law.

10.4. Second Law of Thermodynamics

The Second Law of Thermodynamics highlights a universal tendency toward **disorder**:

Every **energy transfer** increases the **entropy** of the universe.

Living organisms maintain **order** (low entropy) within themselves by releasing waste heat and disorder into their environment. In this way, they comply with the Second Law.

Example: Cells build complex molecules (low entropy) by breaking down glucose, releasing CO_2, H_2O, and heat (high entropy).

10.5. Third Law of Thermodynamics

Although less commonly applied to biology, the Third Law of Thermodynamics states:

As a system approaches **absolute zero** (0 K), the **entropy** of the system approaches a **minimum value**.

This law is mainly relevant to theoretical physics, but it highlights the link between temperature and disorder — even in cells, lower temperatures reduce molecular motion and reaction rates.

10.6. Application to Organisms

Living systems are **open systems**, meaning they exchange both energy and matter with their surroundings. This is how they maintain their organization and function.

Examples of Thermodynamics in Biology

- **Photosynthesis**: Converts light energy into chemical energy, obeying the First Law.
- **Cellular Respiration**: Releases stored chemical energy and produces heat, increasing entropy, following the Second Law.
- **Homeostasis**: Cells regulate temperature, pH, and other factors to optimize **enzymatic activity**, indirectly reflecting thermodynamic

principles.

Thermodynamics shapes every aspect of life, from molecular assembly to energy production and cellular function. Understanding these principles helps explain how organisms power their activities, maintain their internal order, and interact with their environment.

11. ATP as "Fuel"

Every living cell requires energy to function, and the primary energy currency of the **cell** is **adenosine triphosphate**, or **ATP**. This small molecule acts like a rechargeable battery, storing and transferring energy for virtually all cellular processes.

What Makes ATP Special?

ATP consists of:

- An **adenine** base
- A **ribose** sugar
- Three **phosphate groups**

The **phosphate bonds** are high-energy bonds, meaning breaking them releases energy the cell can use for:

- Muscle contraction
- Molecule synthesis
- Transporting ions across membranes

ATP Hydrolysis

The most important **reaction** involving **ATP** is **hydrolysis**:

$$ATP \rightarrow ADP + P_i + energy$$

This release of energy drives most cellular work, linking **catabolic** and **anabolic** pathways.

12. Metabolism and Enzymes in the Cell

Metabolism refers to the sum of all chemical reactions occurring in a cell. It includes both:

- **Catabolic pathways**, which break down molecules to release energy
- **Anabolic pathways**, which build complex molecules, requiring energy input

12.1. Metabolic Complexity

Cells are chemical factories, running thousands of reactions simultaneously. These reactions are tightly linked into pathways, where the product of one step becomes the reactant for the next.

Metabolic Networks

The metabolism of a cell is not linear but forms a complex web of interacting pathways. This allows cells to:

- Adapt to changing environments
- Prioritize essential functions
- Store or release energy efficiently

12.2. Regulation of Enzymes

Enzymes are **biological catalysts** that speed up reactions by lowering the activation energy. Cells carefully regulate their enzymes to control metabolism.

Regulation Mechanisms

- **Allosteric regulation**: Molecules bind to an enzyme at a site other than the active site, changing its activity.

- **Feedback inhibition**: The end product of a pathway inhibits an earlier enzyme to prevent overproduction.
- **Gene regulation**: Cells control enzyme production at the DNA level, turning genes on or off.

12.3. Coenzymes

Coenzymes are organic molecules that assist enzymes in catalyzing reactions. Many vitamins act as coenzymes or precursors.

Examples

- **NAD+** (Nicotinamide adenine dinucleotide) helps transfer electrons in cellular respiration.
- **Coenzyme A** (CoA) helps carry **acetyl groups** in **metabolism**.

12.4. Cofactors

Cofactors are **non-protein helpers** required by some **enzymes** to function. These may be:

- **Metal ions** (like $\mathbf{Mg^{2+}}$ or $\mathbf{Fe^{2+}}$)
- **Small organic molecules** (coenzymes)

Without the right cofactors, many enzymes cannot function.

Together, **enzymes**, **coenzymes**, and **cofactors** ensure that **metabolism** runs smoothly, allowing **cells** to efficiently **convert energy**, build new **molecules**, and respond to **environmental changes**.

13. The Plasma Membrane: The Gatekeeper of the Cell

The plasma membrane is an essential structure that defines the boundary of the cell and controls the movement of substances into and out of the cell. Often called the gatekeeper, it plays a central role in maintaining **cellular homeostasis**.

13.1. Functions of the Plasma Membrane

The plasma membrane serves several critical functions, including:

- Separating the internal environment of the cell from the external environment
- Regulating the import and export of molecules
- Allowing communication between cells
- Anchoring the **cytoskeleton**, providing shape and support
- Participating in cell signaling through receptor proteins

13.2. Structure

The plasma membrane follows the **fluid mosaic model**, where phospholipids, proteins, and carbohydrates constantly shift within a flexible bilayer.

Key Structural Components

- **Phospholipid bilayer** – Two layers of **phospholipids** with **hydrophilic heads** facing outward and **hydrophobic tails** facing inward.
- **Proteins** – Embedded **proteins** act as **channels**, **carriers**, **receptors**, or **enzymes**.
- **Carbohydrates** – Attached to proteins (glycoproteins) or lipids (glycolipids), these aid in cell recognition.

- **Cholesterol** – Maintains fluidity and stability, especially in animal cells.

13.3. Functions

The plasma membrane is far more than just a barrier; it plays a dynamic role in cell function. Key functions include:

- **Selective permeability** – Allowing only certain substances to cross.
- **Transport mechanisms** – Facilitating diffusion, osmosis, facilitated diffusion, and active transport.
- **Signal transduction** – Receiving signals from the environment via receptor proteins, triggering internal responses.
- **Cell-to-cell recognition** – Using **glycoproteins** for immune system responses and cell communication.

13.4. Composition

The **plasma membrane** consists primarily of:

- **Phospholipids** (major structural component)
- **Proteins** (integral and peripheral)
- **Carbohydrates** (cell recognition and adhesion)
- **Cholesterol** (regulates fluidity)

Hydrophobic Tails and Hydrophilic Heads

- **Hydrophilic heads** interact with water inside and outside the cell.
- **Hydrophobic tails** avoid water, forming the inner bilayer core. This dual nature allows the membrane to form a stable barrier in the aqueous environment.

13.5. Proteins

Membrane proteins are essential for transport, communication, and recognition.

Types of Membrane Proteins

- **Transport proteins** – Facilitate movement of molecules across the membrane.
- **Receptor proteins** – Bind signaling molecules to initiate cell responses.
- **Enzymatic proteins** – Catalyze reactions directly at the membrane.
- **Anchor proteins** – Attach to the cytoskeleton for cell shape and stability.

13.6. Cellular Homeostasis

One of the plasma membrane's most important roles is maintaining cellular homeostasis—a stable internal environment despite changing external conditions. This involves:

- Regulating ion concentrations.
- Controlling nutrient uptake.
- Removing waste products.
- Responding to signals from the environment.

Without a **functional plasma membrane**, cells would lose their **integrity** and **die**.

14. Diffusion and Osmosis

Diffusion and osmosis are fundamental processes that describe how molecules and water move across cell membranes. Both are vital for maintaining cellular homeostasis and ensuring proper cell function.

14.1. Diffusion

Diffusion is the **passive movement** of **molecules** from an area of **higher concentration** to an area of **lower concentration**. This movement continues until equilibrium is reached, where molecules are evenly distributed.

Key Features of Diffusion:

- Requires no energy input from the cell.
- Occurs due to random molecular motion.
- Faster in gases than in liquids, and slowest in solids.
- Influenced by temperature, concentration gradient, and molecule size.

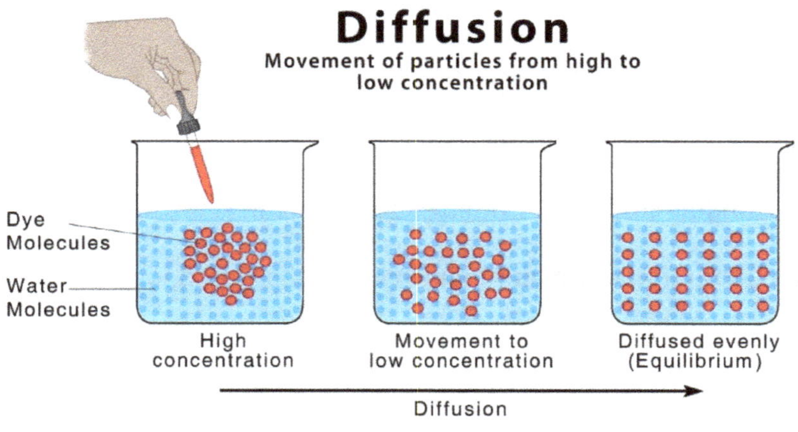

Figure 14.1: Illustration of diffusion process. Source: https://www.sciencefacts.net/diffusion.html

14.2. Osmosis

Osmosis is a special type of diffusion involving the movement of water molecules across a **semipermeable membrane**. Water moves from an area of **lower solute concentration** (more water) to an area of **higher solute concentration** (less water) to balance **concentration differences**.

Key Features of Osmosis:

- Involves only water molecules.
- Requires no energy input.
- Critical for cell volume regulation.
- Can cause cells to swell (**hypotonic solution**), shrink (**hypertonic solution**), or remain the same (isotonic solution).

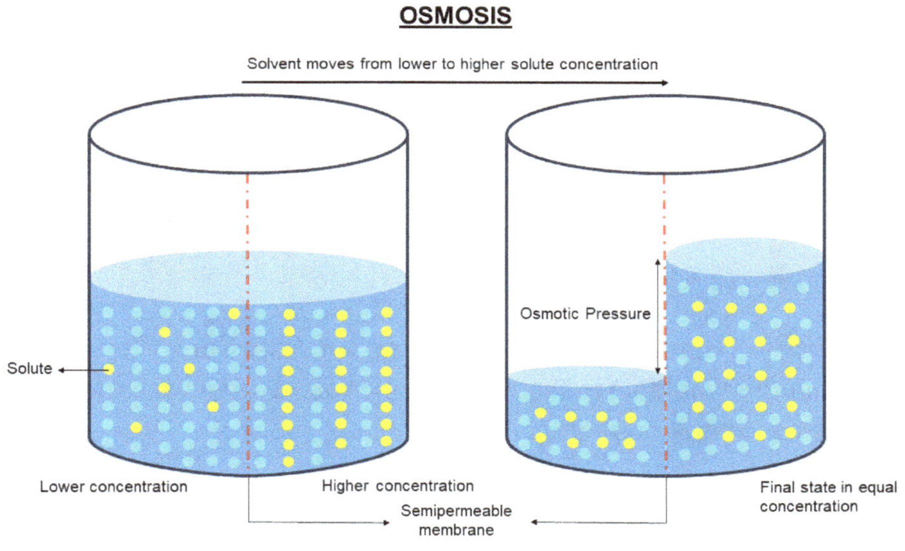

Figure 14.2: Illustration of osmosis process. Source: https://www.biotechreality.com/2023/12/osmosis-definition-diagram-applications-and-faq.html

14.3. Reverse Osmosis: A Process of Osmosis

Reverse osmosis (RO) is the **process** of forcing **water molecules** through a **semipermeable membrane** in the **opposite direction** of **normal osmosis**. This requires the application of external pressure, typically used to purify water.

Key Features of Reverse Osmosis:

- Requires **energy input** (applied pressure).
- Removes **contaminants**, **salts**, and **impurities** from **water**.
- Used in **desalination**, **water purification**, and **wastewater treatment**.

14.4. Similarity to Diffusion

Osmosis and **diffusion** both:

- Involve the **passive movement** of **molecules**.
- Occur without **energy input**.
- Result in **equilibrium** over time.
- Depend on **concentration gradients**.

14.5. Similarity to ODCA

ODCA refers to **Osmotic-Driven Concentration Adjustment** (hypothetical term), the similarity would lie in the way osmotic forces drive water movement to balance solute concentrations across membranes.

- Both processes involve water movement.
- Both adjust solute concentrations to maintain homeostasis.
- Both rely on semipermeable membranes.

14.6. Similarity to DODT

DODT refers to **Diffusion-Oriented Distribution Transport** (another hypothetical term), the similarity lies in how diffusion principles are used to move molecules in biological systems.

- Both rely on **concentration gradients**.
- Both require no **direct energy input**.
- Both help distribute **molecules** evenly across **spaces**.

15. Passive and Active Transport

Cells constantly move **substances** across their **plasma membranes** to maintain **homeostasis** and support **cellular functions**. These **transport processes** fall into two main categories: **passive transport** and **active transport**.

15.1. Processes of Passive Transport

Passive transport is the movement of molecules across the cell membrane without the use of cellular energy (ATP). Instead, molecules move down their concentration gradient — from areas of **high concentration** to areas of **low concentration**.

Main Types of Passive Transport:

- **Simple Diffusion**: Small, non-polar molecules (e.g., oxygen, carbon dioxide) pass directly through the **lipid bilayer**.
- **Facilitated Diffusion**: Larger or polar molecules (e.g., glucose, ions) pass through **channel proteins** or **carrier proteins**.
- **Osmosis**: Movement of water molecules across a **semipermeable membrane**.

15.2. Non-selective Nature of Passive Transport

Passive transport is often non-selective, especially in simple diffusion. Any molecule that fits the physical and chemical properties of the cell membrane (size, polarity, charge) can pass through. This non-selectivity allows cells to continuously exchange essential gases (like oxygen and carbon dioxide) with their environment.

- **Facilitated diffusion**, however, can be more selective due to the specificity of **carrier proteins** and **channel proteins**.
- Non-selectivity benefits processes like **gas exchange**, where all oxygen

molecules can diffuse freely into the cell.

15.3. Effect of Oxygen Content on Passive Transport

The rate of passive transport is heavily influenced by the concentration of oxygen and other molecules. For example, when oxygen concentration outside the cell is high, oxygen molecules diffuse rapidly into the cell. As the oxygen gradient decreases (inside and outside concentrations equalize), the rate of diffusion slows.

Cells in low-oxygen environments depend on steep oxygen gradients to maximize diffusion.

Diffusion rates increase with:

- Higher concentration gradients.
- Smaller molecules.
- Higher temperatures.

15.4. Effect of Metabolic Inhibitors on Active Transport

Unlike passive transport, active transport requires energy in the form of ATP. Active transport moves molecules against their concentration gradient, from **low concentration** to **high concentration** — a process that would not happen spontaneously.

Example:

- The **sodium-potassium pump** actively pumps **sodium ions** out of the **cell** and **potassium ions** into the **cell** — both against their **concentration gradients**.

When **metabolic inhibitors** (like **cyanide**) block ATP production, active transport processes halt. This leads to:

- **Ion imbalances** across the membrane.

- Loss of cellular homeostasis.
- Potential cell death if transport is not restored.

Table 15.1: Comparison of Passive and Active Transport

Feature	Passive Transport	Active Transport
Energy Requirement	None	Requires ATP
Direction	Down concentration gradient	Against concentration gradient
Selectivity	Often non-selective	Highly selective
Rate Limiting Factors	Concentration gradient, temperature	ATP availability, carrier protein function
Examples	Diffusion, osmosis	Sodium-potassium pump, endocytosis

16. Cell Signaling

Cells are not isolated entities — they constantly communicate with their environment and with other cells. This communication process is called cell signaling, which allows cells to coordinate activities, respond to stimuli, and maintain homeostasis. Signals can come from neighboring cells, distant tissues, or even the external environment.

Cell signaling follows a general pathway:

1. **Signal reception** – A signal molecule (ligand) binds to a receptor.
2. **Signal transduction** – The signal is processed inside the cell.
3. **Cellular response** – The cell changes behavior in response to the signal.

16.1. Cell Surface Receptors

Most signals cannot cross the plasma membrane directly, especially if they are hydrophilic or large. Instead, these signals rely on cell surface receptors, proteins embedded in the plasma membrane that detect external signals and trigger internal responses.

Types of Cell Surface Receptors:

- **G-protein coupled receptors (GPCRs)**: These receptors activate intracellular **G-proteins**, which then trigger **secondary messengers** like **cyclic AMP (cAMP)**.
- **Ion channel receptors**: These receptors form **channels** that open when a ligand binds, allowing **ions** to flow across the membrane.
- **Enzyme-linked receptors**: Binding of the signal molecule activates an **enzymatic** domain inside the cell, often triggering **phosphorylation cascades**.

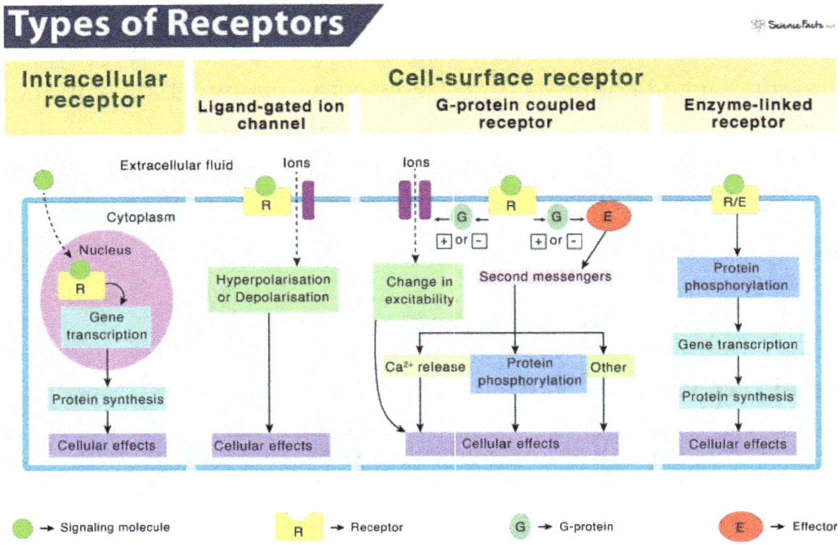

Figure 16.1: Different types of cell surface receptors. Source: https://www.sciencefacts.net/types-of-receptors.html

These **receptors** ensure the cell can detect a wide range of signals, from **hormones** to **neurotransmitters**, and respond appropriately.

16.2. Endocrine Ligands

Some signals travel long distances across the body, such as hormones secreted into the bloodstream. These are called endocrine ligands. Produced by specialized glands, endocrine ligands influence target cells that may be far away from the site of secretion.

Examples of Endocrine Ligands:

- **Insulin**: Secreted by the **pancreas**, insulin regulates **blood glucose levels**.
- **Adrenaline (Epinephrine)**: Released by the **adrenal glands**, it prepares the body for **fight-or-flight** responses.
- **Thyroid hormones**: Influence **metabolism**, **growth**, and **development**.

Endocrine ligands are **stable enough** to survive in the bloodstream and **bind to specific receptors** on target cells, ensuring the right response occurs in the right place.

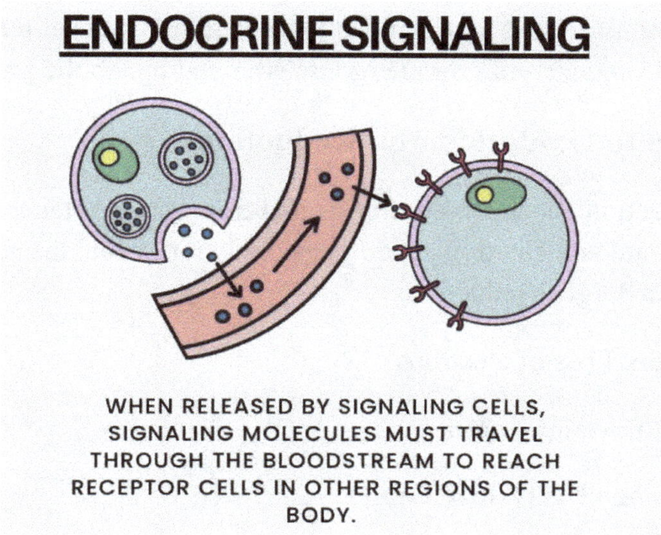

Figure 16.2: Illustration showing endocrine signaling from glands to distant cells. Source: https://medium.com/the-131217net/short-and-long-distance-cell-signaling-ef123e90c8d8

Summary

Cell signaling is essential for **cellular communication**, enabling cells to work together in **multicellular organisms**. Whether through **cell surface receptors** or **endocrine ligands**, cells constantly monitor and respond to internal and external cues to maintain **survival**, **growth**, and **homeostasis**.

17. Oxidation and Reduction

Oxidation and reduction reactions (commonly called **redox reactions**) are fundamental processes in biology, chemistry, and even energy production within living cells. These processes are involved in everything from cellular respiration to photosynthesis and are vital for energy transfer and metabolic processes.

17.1. A Guide to Oxidation and Reduction

In a redox reaction, electrons are transferred from one substance to another. The substance that loses electrons undergoes oxidation, while the substance that gains electrons undergoes reduction.

- **Oxidation**: Loss of electrons.
- **Reduction**: Gain of electrons.

A useful mnemonic is **OIL RIG**:

- **O**xidation **I**s **L**oss (of electrons)
- **R**eduction **I**s **G**ain (of electrons)

These reactions often occur simultaneously, meaning that when one substance is oxidized, another is reduced.

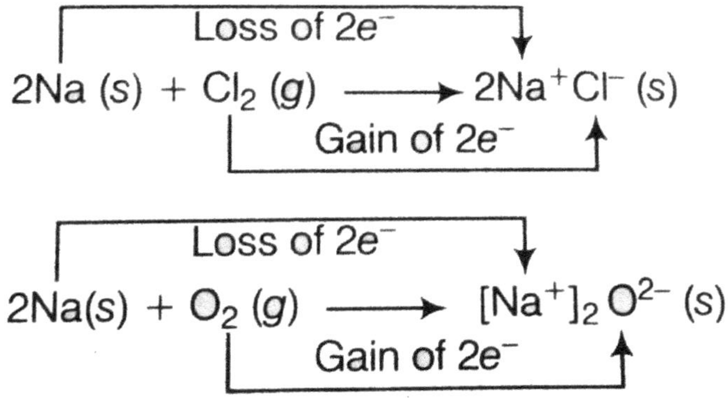

Figure 17.1: Illustration of the electron transfer in a simple redox reaction

17.2. Relationship Between Oxidizing and Reducing Agents

In every redox reaction, there are two key players:

- **Oxidizing agent**: The substance that **accepts electrons** and is **reduced**.
- **Reducing agent**: The substance that **donates electrons** and is **oxidized**.

Example:

In the reaction between sodium and chlorine:

$$2Na + Cl_2 \rightarrow 2NaCl$$

- **Each Sodium (Na)** atom donates an electron (oxidized) — it is the **reducing agent**.
- **Each Chlorine (Cl_2)** atom accepts an electron (reduced) — it is the **oxidizing agent**.

These agents work together to drive the overall redox process.

17.3. Nature of Redox Reactions

Redox reactions are crucial in **biology** because they are involved in:

- **Cellular respiration**, where glucose is oxidized to produce **ATP**.
- **Photosynthesis**, where water is split (oxidized) and carbon dioxide is reduced.
- **Detoxification processes** in cells, where harmful substances are **neutralized** by redox mechanisms.

Redox reactions also play a key role in industrial processes, such as:

- **Combustion** in engines.
- **Corrosion** of metals.
- **Batteries** and **fuel cells**.

17.4. Methods of Redox Reactions

There are several ways to classify and balance redox reactions, particularly in **aqueous solutions**:

- **Half-reaction method**: Splitting the overall reaction into **oxidation** and **reduction** halves.
- **Ion-electron method**: Tracking electrons explicitly.
- **Oxidation number method**: Assigning oxidation numbers to atoms and observing changes.

Example Using Half-Reactions

$$Zn + Cu^{2+} \rightarrow Zn^{2+} + Cu$$

Oxidation half-reaction:

$$Zn \rightarrow Zn^{2+} + 2e^-$$

Reduction half-reaction:

$$Cu^{2+} + 2e^- \rightarrow Cu$$

17.5. Examples of Redox Reactions

Biological Example: Cellular Respiration

Glucose is **oxidized**, and oxygen is **reduced**:

$$C_6H_{12}O_6 + 6O_2 \rightarrow 6CO_2 + 6H_2O$$

Industrial Example: Rust Formation

Iron reacts with oxygen in the presence of water:

$$4Fe + 3O_2 + 6H_2O \rightarrow 4Fe(OH)_3$$

- **Iron (Fe)** is **oxidized** to **$Fe(OH)_3$**.
- **Oxygen (O_2)** is **reduced** to form hydroxide.

Summary

Redox reactions are a core part of **biological processes** and **chemical transformations**. From **energy production** in cells to **environmental processes** like rusting and combustion, understanding the flow of **electrons** helps explain the behavior of materials and living systems alike.

18. Steps of Cellular Respiration

Cellular respiration is the process by which cells extract energy from glucose and store it in the form of ATP (adenosine triphosphate). This is a **multi-step**, carefully regulated process that takes place in both the **cytoplasm** and the **mitochondria** of cells.

The overall equation for **cellular respiration** is:

$$C_6H_{12}O_6 + 6O_2 \rightarrow 6CO_2 + 6H_2O + Energy(ATP)$$

18.1. ATP – The Energy Currency

ATP acts as the main energy carrier within the cell. It stores energy in phosphate bonds, which can be broken to release energy for cellular work, such as muscle contraction, active transport, and biosynthesis.

Key Features of ATP

- Composed of adenine, ribose, and three phosphate groups.
- The bond between the second and third phosphate group is where energy is stored.
- When ATP loses a phosphate group, it becomes **ADP** (adenosine diphosphate) and **releases energy**.

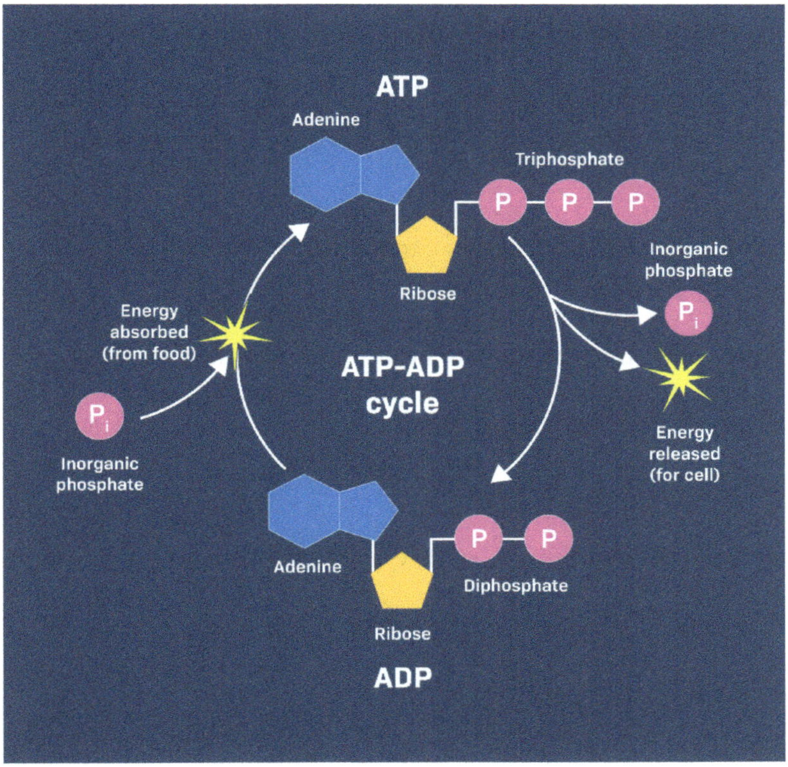

Figure 18.1: ATP recycling process. Source: https://www.labxchange.org/library/items/lb:LabXchange:8ef89bac:html:1

18.2. NADH – The Electron Shuttle

NADH (nicotinamide adenine dinucleotide) is a vital electron carrier in cellular respiration. During various stages of respiration, electrons are stripped from glucose and transferred to **NAD+**, converting it to **NADH**.

- NADH carries high-energy **electrons** to the **electron transport chain**, where their energy is used to produce **ATP**.
- Another electron carrier, **FADH2**, plays a similar role.

18.3. CO_2 – The Waste Product

During cellular respiration, carbon atoms from glucose are released as carbon dioxide (CO_2). This occurs mainly in the **Krebs cycle** (also called the **citric acid cycle**).

- In **aerobic organisms**, CO_2 is expelled from the cell and ultimately exhaled by the lungs (in animals).
- In **plants**, CO_2 may be recycled during photosynthesis.

Key Sources of CO_2 in Respiration:

- **Glycolysis** produces no CO_2.
- **Pyruvate oxidation** releases 1 CO_2 per pyruvate.
- **Krebs cycle** releases 2 CO_2 per turn.

18.4. Glucose – The Starting Point

Glucose ($C_6H_{12}O_6$) is the primary **fuel** for cellular respiration. Cells break down glucose gradually, extracting electrons and chemical energy at every step.

Steps of Glucose Breakdown:

1. **Glycolysis** (in cytoplasm) – Splits glucose into two **pyruvate** molecules.
2. **Pyruvate oxidation** (in mitochondria) – Converts pyruvate into **acetyl-CoA**.
3. **Krebs cycle** – Fully breaks down acetyl-CoA, releasing **CO_2**, **NADH**, and **FADH2**.
4. **Electron Transport Chain (ETC)** – Uses NADH and FADH2 to drive ATP production.

Energy Yield:

- **Glycolysis**: 2 ATP
- **Krebs Cycle**: 2 ATP
- **Electron Transport Chain**: ~28-32 ATP

Summary

Cellular respiration is the essential process that allows cells to extract energy from **glucose** and store it in **ATP**, the cell's primary energy carrier. **NADH** acts as a vital **electron shuttle**, and **CO_2** is produced as a waste product. Understanding these steps helps explain how living things power all their activities, from simple **growth** to complex **movement** and **communication**.

19. Introduction to Photosynthesis

Photosynthesis is the remarkable biological process by which **plants**, **algae**, and some **bacteria** harness **light energy** to produce **chemical energy**. It occurs mainly in the **chloroplasts** of **plant cells**, converting **carbon dioxide (CO_2)** and **water (H_2O)** into **glucose** and **oxygen (O_2)**. This process fuels the food chain and supports almost all life on Earth.

The general equation for **photosynthesis** is:

$$6CO_2 + 6H_2O + \text{light energy} \rightarrow C_6H_{12}O_6 + 6O_2$$

19.1. Light Energy

The energy source driving photosynthesis is light energy from the **sun**. Plants use **chlorophyll**, a green **pigment** found in **chloroplasts**, to absorb this light, primarily in the **red** and blue wavelengths, while reflecting green light.

Key Points:

- **Photons** are packets of **light energy**.
- Light energy excites **electrons** in chlorophyll, initiating the process.
- Without light, **photosynthesis** cannot proceed.

19.2. Reactants

The raw materials required for **photosynthesis** are:

- **Carbon dioxide (CO_2)** from the atmosphere, taken in through **stomata**.
- **Water (H_2O)** absorbed from the soil by **roots**.
- **Light energy** from the **sun**.

These three reactants come together in the chloroplast, where photosynthesis occurs.

19.3. Light-dependent Reactions

The first stage of **photosynthesis** is the **light-dependent reactions**, which take place in the **thylakoid membranes** inside the **chloroplast**.

Key Events:

- **Chlorophyll** absorbs light energy, exciting electrons.
- **Water molecules** are split (photolysis), releasing oxygen (O_2), protons (H^+), and electrons.
- **ATP** and **NADPH** are produced, carrying energy and electrons to the next stage.

Summary Reaction:

$$H_2O + Light\ energy \rightarrow ATP + NADPH + O_2$$

19.4. Reactions of the Calvin Cycle

The second stage, called the **Calvin Cycle** (or **light-independent reactions**), occurs in the **stroma** of the **chloroplast**. This process uses the products of the **light-dependent reactions** — **ATP** and **NADPH** — along with **carbon dioxide** to **synthesize glucose**.

Key Steps:

- **Carbon fixation** (incorporating CO_2 into organic molecules).
- **Reduction phase**, producing **G3P**, a sugar precursor.
- **Regeneration** of the molecule **RuBP**, enabling the cycle to continue.

Summary Reaction:

$$CO_2 + ATP + NADPH \rightarrow C_6H_{12}O_6$$

19.5. Carbon Fixation

Carbon fixation is the process of **converting inorganic carbon** (CO_2) into **organic molecules** like **glucose**. It occurs during the **Calvin Cycle** when **carbon dioxide** is attached to a **five-carbon compound** called **RuBP**. This step is **catalyzed** by the enzyme **RuBisCO** — the most abundant enzyme on Earth.

19.6. Carbon Dioxide

Carbon dioxide (CO_2) is a key reactant in photosynthesis. It diffuses into leaves through **stomata**, tiny openings on the **leaf surface**. CO_2 levels, temperature, and light intensity all influence the **rate of photosynthesis**.

Importance:

- Provides the **carbon atoms** used to form **glucose**.
- Increasing atmospheric **CO_2** (due to human activity) can **enhance photosynthesis** — a phenomenon called **CO_2 fertilization**.

Summary

Photosynthesis is vital for all life on Earth, as it transforms **solar energy** into the **chemical energy** stored in **glucose**, while simultaneously producing the **oxygen** we breathe. By understanding **light-dependent reactions**, the **Calvin Cycle**, and **carbon fixation**, we appreciate how plants and photosynthetic organisms sustain the **biosphere**.

20. Cytoskeleton

The cytoskeleton is a dynamic network of protein filaments that provides structural support, shape, and organization to cells. It acts like the scaffolding of a building, holding everything in place while also enabling essential cellular movements and transport processes. The cytoskeleton is vital in cell division, intracellular transport, and even cell signaling.

20.1. Microtubules

Microtubules are hollow tubes made up of tubulin proteins. They are the largest filaments in the cytoskeleton and play key roles in:

- **Cell shape maintenance**
- **Intracellular transport** (acting like tracks for motor proteins like **kinesin** and **dynein**)
- **Chromosome separation** during cell division
- **Organelle positioning**

Microtubules are highly dynamic, constantly assembling and disassembling in response to cellular needs — a process known as **dynamic instability**.

20.2. Intermediate Filaments

Intermediate filaments are fibrous proteins that form rope-like fibers, giving cells mechanical strength. They are less dynamic than microtubules or microfilaments but are crucial for maintaining the integrity of the cytoskeleton.

Examples include:

- **Keratin** (in epithelial cells)
- **Lamin** (which supports the nuclear envelope)

Intermediate filaments are essential for withstanding **mechanical stress**, especially in tissues like skin, muscles, and nerves.

20.3. Microfilaments

Microfilaments, also known as actin filaments, are the thinnest components of the cytoskeleton. Made of **actin proteins**, they are responsible for:

- **Cell shape changes**
- **Amoeboid movement** (cell crawling)
- **Cytokinesis** (division of the cytoplasm during cell division)
- **Intracellular transport of small organelles and vesicles**

Microfilaments work closely with **myosin** motor proteins to produce movement, particularly in **muscle cells**.

20.4. Spindle Apparatus

The spindle apparatus is a **microtubule-based structure** formed during cell division. It is responsible for:

- **Aligning chromosomes** at the cell's equator during **metaphase**
- **Separating sister chromatids** during **anaphase**

The spindle originates from **centrosomes** in animal cells, which act as **microtubule-organizing centers** (MTOCs).

20.5. Functions of the Cytoskeleton

The cytoskeleton has **multifunctional roles**, including:

- Maintaining **cell shape** and **internal organization**
- Facilitating **intracellular transport** of organelles, vesicles, and molecules

- Enabling **cell migration**, division, and intracellular signaling
- Providing mechanical **resistance to deformation**
- **Connecting** cells to external structures via the **extracellular matrix** (ECM)

20.6. Structure of the Cytoskeleton

The cytoskeleton is a **highly organized yet flexible network**, constantly remodeling to meet the **cell's needs**. The three main components — **microtubules**, **intermediate filaments**, and **microfilaments** — work together in a **coordinated manner**, often interacting with **motor proteins**, **cell membranes**, and other cellular structures.

20.7. Research and Medical Importance

The **cytoskeleton** is a critical focus of **biomedical research** because it plays a key role in:

- **Cancer metastasis** (how cancer cells move)
- **Neurodegenerative diseases** (defective cytoskeletal proteins can impair **neuronal transport**)
- **Pathogen infection** (some viruses and bacteria hijack the cytoskeleton to invade cells)
- **Drug development** (certain chemotherapy drugs, like **taxanes**, target microtubules to halt cancer cell division)

Ongoing research continues to uncover new roles of the cytoskeleton in stem cell biology, tissue regeneration, and cellular aging.

21. How Cells Move

Cells are not static entities — they move, change shape, and migrate in response to their environment. **Cell movement** is essential for processes like **wound healing**, **immune response**, **development**, and even **cancer metastasis**. Cells achieve movement through a highly coordinated interaction between their **cytoskeleton**, **cell membrane**, and **extracellular matrix (ECM)**.

21.1. Focal Adhesions

Focal adhesions are specialized sites where cells attach to the extracellular matrix. These adhesive points act like anchor points, allowing the cell to push or pull itself forward.

They consist of:

- **Integrin proteins** that span the plasma membrane
- **Cytoskeletal proteins** (like **actin** and **talin**) that connect to the internal cytoskeleton
- **Signaling molecules** that regulate attachment and detachment

The dynamic assembly and disassembly of **focal adhesions** allow cells to **crawl** across surfaces.

21.2. Lamellipodia

Lamellipodia are sheet-like protrusions formed at the leading edge of a migrating cell. They are rich in actin filaments, which polymerize rapidly to push the plasma membrane outward.

The formation of lamellipodia helps cells:

- Probe their environment
- Extend their front edge

- Grip onto the **ECM** via **focal adhesions**

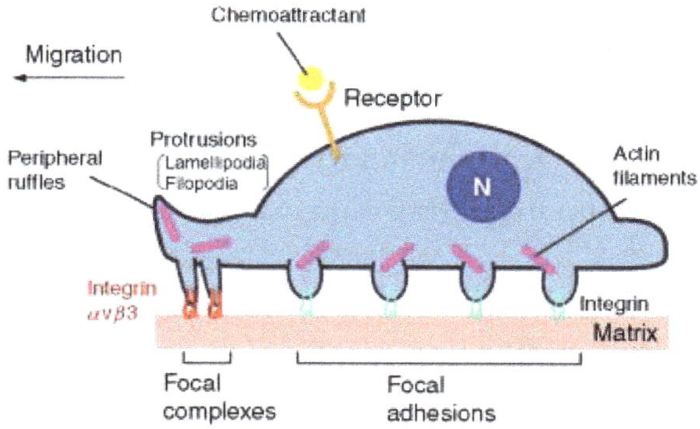

Figure 21.1: A migrating cell with visible lamellipodia.Source: https://www.sciencedirect.com/topics/immunology-and-microbiology/cell-migration

21.3. Actin Filaments in Cell Movement

Actin filaments play a central role in cell movement. They form branched networks that push the cell membrane outward in **lamellipodia**, while **contractile bundles** of actin at the rear of the cell help **retract** the trailing edge.

Key processes involving actin:

- **Polymerization** at the front of the cell
- **Cross-linking** to form a gel-like mesh
- **Myosin motor proteins** contracting actin bundles to pull the cell body forward

Actin filaments also interact with **microtubules** to guide **directional movement**.

21.4. Cellular Transport and Movement Coordination

While cytoskeletal elements drive the mechanical aspect of movement, **intracellular transport** ensures proper **positioning** of organelles, vesicles, and signaling molecules necessary for directed movement.

- **Microtubules** act as highways, transporting signaling molecules to the cell front.
- **Motor proteins**, like **kinesin** and **dynein**, transport **vesicles** along microtubules to support movement and adhesion.

This **interplay between transport and cytoskeletal dynamics** ensures coordinated, purposeful movement.

21.5. Speed of Cell Movement

Different types of cells move at different speeds depending on their function and environment:

- **Fibroblasts** (in wound healing) move at about **1-2 microns per minute**.
- **Neutrophils** (immune cells) can move up to **20 microns per minute**.
- **Cancer cells** that metastasize can vary significantly depending on tissue stiffness and molecular signaling.

Factors affecting speed:

- Surface adhesion strength
- Availability of ECM attachment points
- Actin polymerization rate
- External chemical signals (chemotaxis)

Cells strike a balance between **speed** and **stability** — too much adhesion slows movement, while too little prevents proper traction.

Table 21.1: Comparative Speeds of Different Cell Types in Migration

Cell Type	Average Speed (microns per minute)	Notes
Fibroblasts	1 - 2	Important in wound healing and tissue repair; slow and steady migration.
Neutrophils	10 - 20	Fast-moving immune cells that chase pathogens during infection.
Epithelial Cells	~0.5 - 1	Usually part of epithelial sheets; move slowly during wound closure.
Endothelial Cells	~1 - 2	Line blood vessels; migrate during angiogenesis.
Mesenchymal Stem Cells	2 - 5	Move during development and tissue repair.
Cancer Cells (Metastatic)	Variable (0.1 - 10+)	Speed depends heavily on tissue type, stiffness, and local signaling.
Keratinocytes	1 - 3	Skin cells that migrate during wound healing.
Amoebae (Dictyostelium)	5 - 10	Model organism for studying cell movement.

Summary

The ability of cells to move is critical for **development**, **immune surveillance**, **tissue repair**, and even **disease progression**. This movement is orchestrated by the **cytoskeleton**, **adhesion proteins**, and **intracellular signaling pathways**, all working together to allow cells to respond to their environment and perform their biological functions.

22. Cellular Digestion

The process by which cells break down **macromolecules**, recycle cellular components, and manage waste disposal is collectively known as **cellular digestion**. This process is essential for maintaining **cellular homeostasis**, supporting **energy generation**, and ensuring the removal of potentially harmful waste products. Cellular digestion occurs in **organelles** specially adapted for this purpose, particularly **lysosomes** and **vacuoles**.

22.1. Intracellular Digestion

Intracellular digestion refers to the breakdown of **materials** within the cell. It typically occurs in the following steps:

1. **Endocytosis**: The cell engulfs external particles (such as nutrients or pathogens) through the formation of a **vesicle**.
2. **Fusion with Lysosome**: The vesicle fuses with a **lysosome**, a specialized organelle packed with **digestive enzymes**.
3. **Enzymatic Breakdown**: Enzymes within the lysosome break down the **proteins**, **lipids**, **carbohydrates**, and **nucleic acids** into smaller molecules that can be reused by the cell.
4. **Nutrient Recycling**: The digested products are transported into the cytoplasm for cellular use.
5. **Waste Elimination**: Indigestible or harmful materials are packaged into **exocytotic vesicles** and expelled from the cell.

The process can also occur for internal cellular components through **autophagy**, where the cell digests its own worn-out organelles to recycle materials and maintain cell health.

22.2 Functions of Digestive Enzymes

Digestive enzymes play a critical role in intracellular digestion. These enzymes belong to several families:

Table 22.1: Types of Digestive Enzymes and Their Functions in Intracellular Digestion

Enzyme Type	**Substrates**	**Function**
Proteases	Proteins	Break down proteins into amino acids
Lipases	Lipids	Decompose fats into fatty acids and glycerol
Nucleases	DNA, RNA	Degrade nucleic acids into nucleotides
Carbohydrases	Carbohydrates	Break down polysaccharides into sugars

These enzymes operate optimally in the acidic environment of the lysosome, ensuring that digestion is compartmentalized and does not harm the rest of the cell.

Table 22.2: Types of Digestive Enzymes and Their Substrates

Digestive Enzyme	Substrate	Primary Function
Proteases (Peptidases)	Proteins	Break down proteins into smaller peptides and amino acids
Amylases	Starch (Carbohydrates)	Break down starch into maltose and simple sugars
Lipases	Fats (Lipids)	Break down triglycerides into glycerol and free fatty acids
Nucleases	DNA, RNA (Nucleic Acids)	Break down DNA and RNA into nucleotides
Phosphatases	Phosphate-containing molecules	Remove phosphate groups from molecules
Lysozymes	Bacterial cell walls	Break down peptidoglycan walls of bacteria

22.3 Pathogenesis of Diseases Caused by Cellular Digestion

Disruptions in **cellular digestion** can result in **serious diseases**. These conditions are often linked to **lysosomal storage disorders**, where the lysosome is unable to break down certain substances, leading to their accumulation and eventual cell damage. Some examples include:

- **Gaucher Disease**: Caused by a deficiency in **glucocerebrosidase**, leading to the accumulation of **fatty substances** in cells, particularly in the liver, spleen, and bone marrow.
- **Tay-Sachs Disease**: Results from a lack of **hexosaminidase A**, which causes toxic buildup of **gangliosides** in neurons, ultimately leading to

neurodegeneration.

- **Pompe Disease**: Due to defects in **acid alpha-glucosidase**, leading to the buildup of **glycogen** in muscle cells.

These diseases illustrate the critical importance of proper lysosomal function for cell survival and health.

Chapter Summary

- **Cellular digestion** is essential for recycling nutrients and removing waste.
- **Intracellular digestion** occurs within **lysosomes**, with the help of specialized **enzymes**.
- Disruptions in digestion lead to **lysosomal storage diseases**, some of which have **severe** clinical consequences.

23. What is Genetic Material?

Life depends on the storage, transmission, and expression of **genetic material** — the biological information that determines the structure, function, and behavior of every living organism. Understanding the nature and components of genetic material helps us comprehend how traits are inherited, how evolution occurs, and how diseases develop. This chapter explores the molecules and structures that make up genetic material and how they contribute to the flow of genetic information in cells.

23.1. DNA – The Blueprint of Life

DNA (Deoxyribonucleic Acid) is the primary carrier of **genetic information** in most living organisms. It contains the **instructions** needed for the development, functioning, and reproduction of all cells.

Key Features of DNA:

- Double-helix structure with two strands running in opposite directions.
- Composed of **nucleotides**, each consisting of a **sugar (deoxyribose)**, a **phosphate group**, and a **nitrogenous base** (Adenine, Thymine, Cytosine, or Guanine).
- **Base-pairing rules**: Adenine pairs with Thymine, Cytosine pairs with Guanine.
- Encodes information through the **sequence of bases**, which acts like a biological code.

Functions of DNA

- Stores **genetic information**.
- Directs the **synthesis of proteins** through an intermediary process called **transcription**.

- Ensures accurate **replication** during cell division.
- Allows **evolutionary changes** through mutations.

23.2. RNA – The Messenger and More

RNA (Ribonucleic Acid) is another vital **nucleic acid** that plays multiple roles in cells. Unlike DNA, **RNA** is usually single-stranded, and it contains **ribose** sugar and **Uracil (U)** instead of Thymine.

Types of RNA:

- **mRNA (Messenger RNA)** – Carries genetic instructions from **DNA** to the **ribosome** for **protein synthesis**.
- **tRNA (Transfer RNA)** – Brings the appropriate **amino acids** to the ribosome during **translation**.
- **rRNA (Ribosomal RNA)** – Structural and functional component of the **ribosome**.

Functions of RNA

- Converts **genetic instructions** into proteins.
- In some viruses (like retroviruses), **RNA** acts as the primary genetic material.
- Plays regulatory roles, such as in **RNA interference (RNAi)**.

23.3. Genes – Units of Heredity

A **gene** is a specific segment of **DNA** that contains the **instructions** for building a particular protein or functional RNA molecule.

Key Features of Genes:

- **Each gene** has a **promoter region**, coding region, and regulatory sequences.

- The **genetic code** (triplet codons) determines the sequence of **amino acids** in a protein.
- Genes can undergo **mutations**, leading to changes in protein structure and function.
- Genes are organized on **chromosomes**.

Functions of Genes

- Carry **instructions for traits**.
- Allow cells to respond to **environmental changes**.
- **Regulate development** and cellular processes.

23.4. Chromosomes – Packages of Genetic Information

Chromosomes are **highly organized structures** that contain long strands of **DNA** wrapped around **histone proteins**. Every organism has a characteristic number of chromosomes.

Types of Chromosomes:

- **Prokaryotic Chromosome** – A single circular DNA molecule.
- **Eukaryotic Chromosomes** – Linear DNA molecules located inside the **nucleus**, often occurring in **pairs**.

Functions of Chromosomes

- **Ensure accurate DNA replication** and distribution during cell division.
- **Protect genetic material** from damage.
- Facilitate **gene regulation** by controlling accessibility to the DNA.

23.5. Non-coding DNA – More Than Junk

Although **genes** are essential, a significant portion of the genome does not code for proteins. This **non-coding DNA** plays **regulatory, structural, and evolutionary roles**.

Functions of Non-coding DNA:

- Contains **regulatory sequences** (enhancers, silencers) that control **gene expression**.
- Provides sites for the **attachment of proteins** involved in chromosome structure and function.
- May serve as raw material for the evolution of **new genes**.
- Some non-coding regions contain **transposable elements**, which can **move** within the genome, affecting genetic diversity.

Figure 23.1 shows important differences between DNA and RNA.

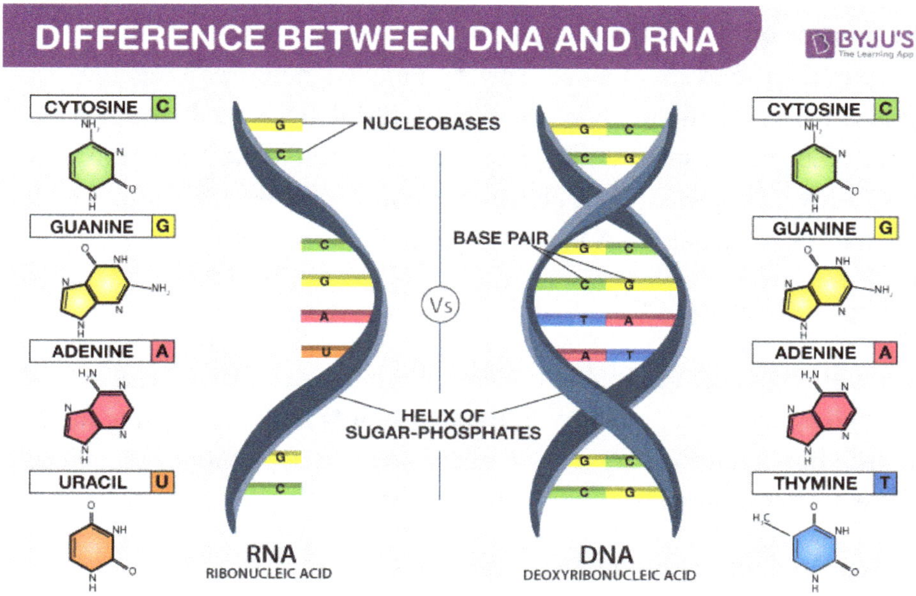

Figure 23.1: Important differences between DNA and RNA. Source: https://byjus.com/biology/difference-between-dna-and-rna/

Table 23.1: Structure of DNA and RNA with Base Pairing and Nucleotide Components

Feature	DNA	RNA
Strand Type	Double-stranded helix	Single-stranded
Sugar	Deoxyribose	Ribose
Bases	Adenine (A), Thymine (T), Cytosine (C), Guanine (G)	Adenine (A), **Uracil (U)**, Cytosine (C), Guanine (G)
Base Pairing	A-T, C-G	A-U, C-G
Overall Structure	Double helix with complementary base pairing	Single strand folding into various shapes
Location	Nucleus (Eukaryotes), Cytoplasm (Prokaryotes)	Cytoplasm (all cells)
Function	Genetic information storage	Protein synthesis and regulatory roles

24. The Replication of DNA

The process by which **DNA** copies itself is known as **DNA replication** — a vital step in the life cycle of a cell, ensuring that each new cell receives an accurate copy of genetic material. Without faithful **DNA replication**, cells would not inherit the correct instructions necessary for survival and function. This process follows the principle of **complementary base pairing**, allowing the sequence of the original DNA strand to guide the formation of its exact copy.

24.1. Base Pairing

Base pairing is central to the replication process. The **two strands** of the **DNA double helix** are held together by **hydrogen bonds** between complementary **nucleotide bases**. During replication, these **strands separate**, and each acts as a **template** for building a new complementary strand.

The **base pairing rules** guide this process:

- **Adenine (A)** pairs with **Thymine (T)**
- **Cytosine (C)** pairs with **Guanine (G)**

This complementary pairing ensures that each new DNA strand is an **exact replica** of the original.

24.2. Linear Replication

In **eukaryotic cells**, DNA replication occurs along **linear chromosomes**. This process starts at **multiple origins of replication** along the chromosome to speed up the copying process.

Steps in Linear Replication:

1. **Unwinding:** DNA **helicase** separates the two strands by breaking **hydrogen bonds**.

2. Priming: An **RNA primer** is added by **primase**, which acts as a starting point for synthesis.

3. Elongation: DNA polymerase adds complementary **nucleotides** to the exposed bases, building the new strand.

4. Leading vs. Lagging Strand:

The **leading strand** is synthesized continuously.

The **lagging strand** is synthesized in small pieces called **Okazaki fragments**, which are later joined together.

5. Proofreading: DNA polymerase checks for errors and corrects mismatches.

6. Completion: The new strands are sealed by **DNA ligase**, completing two identical **DNA molecules**.

24.3. Self-replication

The remarkable ability of **DNA** to copy itself precisely stems from its **self-replicating nature**. This means that **each strand serves as a template**, guiding the formation of a new complementary strand. This is known as **semi-conservative replication**, meaning that each daughter DNA molecule contains **one old (template) strand and one newly synthesized strand**.

This **self-replicating property** ensures:

- Genetic information is faithfully transmitted from **cell to cell** during **cell division**.
- Genetic information is passed from **parent to offspring**, ensuring continuity of life.

Key Concept:
The structure of DNA — specifically, the **complementary base pairing** and **anti-**

parallel strands — is the key feature that allows for **self-replication** with minimal error.

Key Takeaways

- **DNA replication** ensures that cells inherit accurate copies of genetic material.
- **Complementary base pairing** is the foundation of the replication process.
- **Linear replication** occurs along chromosomes in eukaryotic cells, with both leading and lagging strands.
- **Self-replication** allows DNA to act as both the carrier of genetic information and the template for its own duplication.

25. What is Cell Reproduction?

Cell reproduction is the biological process through which cells create **new cells**, ensuring the continuity of life and the proper growth and maintenance of organisms. It plays a crucial role in everything from **development** and **healing** to **reproduction** and **evolution**. All living organisms — from single-celled bacteria to complex multicellular species — depend on this process to sustain life and pass on genetic information to future generations.

25.1. Basic Steps in Cell Reproduction

Cell reproduction generally follows a set sequence of events, ensuring that each **daughter cell** receives a complete set of **genetic instructions** and sufficient **cellular components** to function. The key steps include:

1. **DNA Replication:** Before a cell divides, it must copy its entire genome so each daughter cell has a complete set of genetic instructions.
2. **Organelle Duplication:** In eukaryotic cells, essential organelles (like mitochondria) also replicate to provide each daughter cell with the machinery it needs.
3. **Segregation of Genetic Material:** The duplicated **chromosomes** are sorted and distributed into the new cells.
4. **Division of Cytoplasm:** The cell's cytoplasm is split between the two new cells.
5. **Cell Membrane Formation:** In animal cells, the **cell membrane** pinches off, forming two distinct cells. In plants, a **cell plate** forms between the two cells.

25.2. Cell Division

Cell division is the physical process where one cell splits into **two daughter cells**. There are two major types:

- **Mitosis:** Produces two **genetically identical** daughter cells. This occurs in **somatic cells** for growth, repair, and maintenance.
- **Meiosis:** Produces **four genetically unique** daughter cells with half the genetic material (haploid). This process creates **gametes** (sperm and egg cells) for sexual reproduction.

Both processes are tightly controlled to prevent **errors**, such as **incomplete genetic copies** or **uncontrolled growth**.

Table 25.1: Comparison Between Mitosis and Meiosis

Feature	Mitosis	Meiosis
Purpose	Growth, tissue repair, and asexual reproduction	Formation of gametes for sexual reproduction
Location	Occurs in somatic (body) cells	Occurs in reproductive cells (germ cells)
Number of Divisions	1 cell division	2 consecutive cell divisions
Number of Daughter Cells	2 daughter cells	4 daughter cells
Genetic Makeup of Daughter Cells	Genetically identical to parent cell	Genetically unique from parent and each other
Chromosome Number in Daughter Cells	Diploid (same as parent)	Haploid (half of parent's chromosome number)
Crossing Over	Does not occur	Occurs during Prophase I
Independent Assortment	Does not occur	Occurs during Metaphase I
Function in Multicellular Organisms	Growth, repair, maintenance	Production of gametes
Example Organism	Skin cell division	Formation of sperm and egg cells

25.3. Meiosis

Meiosis is a specialized type of **cell division** that reduces the **chromosome number** by half, producing **gametes** (sperm or eggs). It consists of two consecutive divisions:

- **Meiosis I:** Homologous chromosomes pair up and exchange genetic material (crossing over), then segregate into different cells.
- **Meiosis II:** Sister chromatids separate into individual cells, resulting in four genetically unique haploid cells.

This **genetic reshuffling** introduces **variation**, which is the foundation for **evolution**.

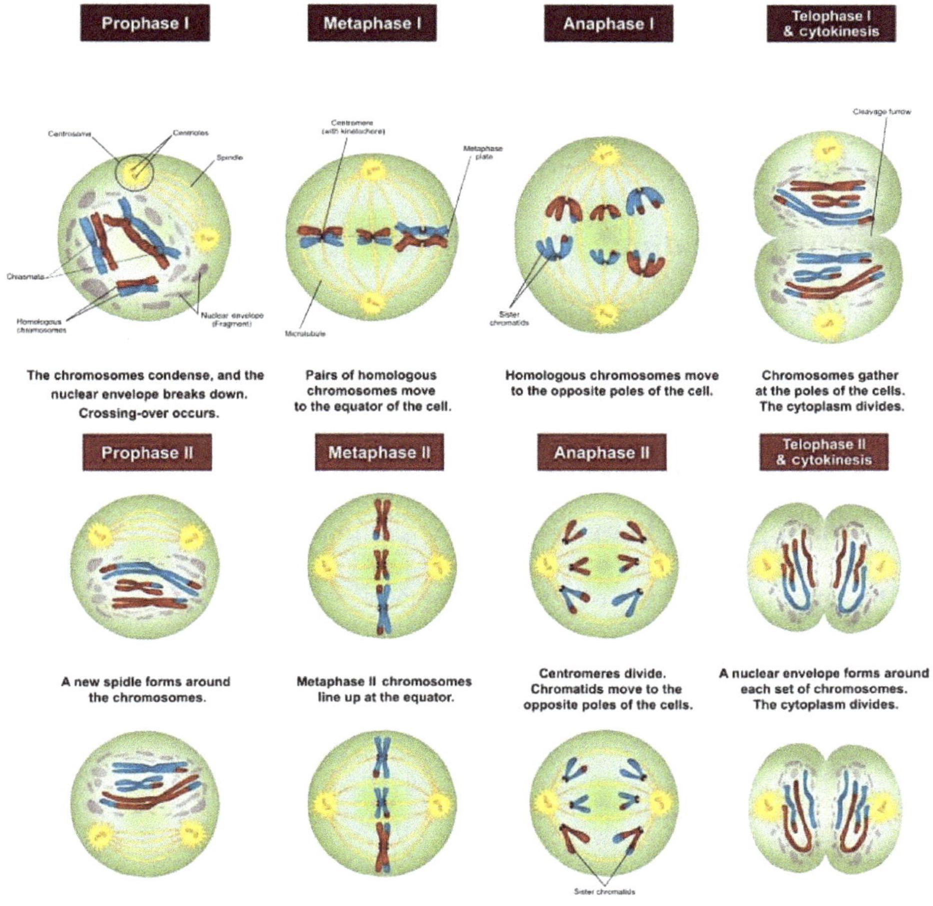

Figure 25.1: Stages of meiosis. Source: https://opengenetics.pressbooks.tru.ca/chapter/meiosis-i/

25.4. Cytokinesis

Cytokinesis is the final step in cell division where the **cytoplasm** divides, creating two distinct cells. The method varies between **animal cells** and **plant cells**:

- **Animal cells:** The **plasma membrane** pinches inward, dividing the cytoplasm (a process called the **cleavage furrow**).
- **Plant cells:** A **cell plate** forms between the daughter cells, eventually becoming a new **cell wall**.

Cytokinesis ensures that each daughter cell receives an equal share of **cytoplasm** and organelles.

25.5. Gamete Formation

In sexually reproducing organisms, **gametes** (sperm and egg cells) are formed through **meiosis**. These cells contain **half the number of chromosomes** (haploid), so when they fuse during fertilization, the **diploid** number is restored.

Key features of **gamete formation**:

- **Spermatogenesis:** Formation of sperm in males.
- **Oogenesis:** Formation of eggs in females.
- The process allows for **genetic diversity** through **independent assortment** and **crossing over**.

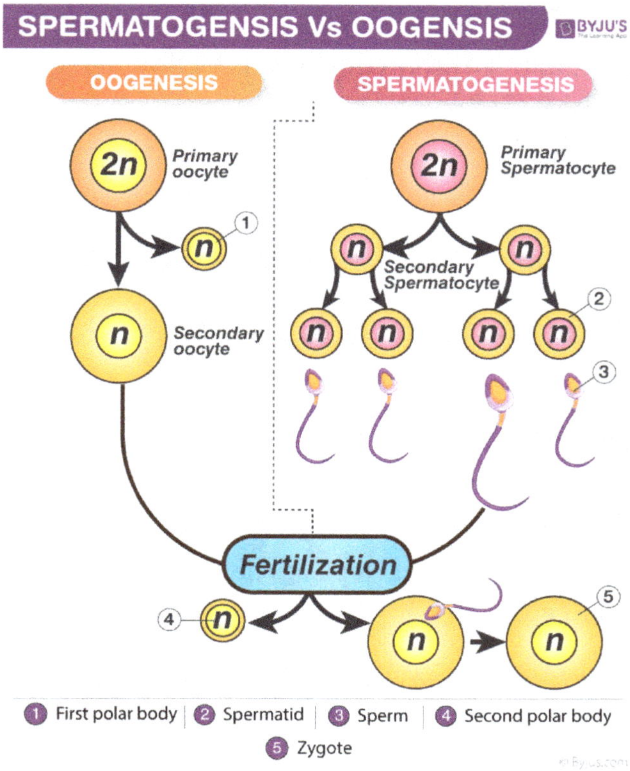

Figure 25.2: Difference between spermatogenesis and oogenesis. Source: https://byjus.com/biology/difference-between-spermatogenesis-and-oogenesis/

25.6. Chemical Replication

Beyond physical division, **chemical replication** is the core biological process where **nucleic acids**, **proteins**, and other biomolecules are synthesized to ensure new cells can function. This replication is guided by the **genetic code** stored in **DNA** and transcribed into **RNA**, then translated into **proteins** — the workhorses of the cell.

25.7. Leukemia Case Study

Uncontrolled **cell reproduction** can lead to diseases such as **cancer**. One notable example is **leukemia**, a type of **blood cancer** where **white blood cells** proliferate uncontrollably. This disrupts the normal balance of **blood cells**, leading to **anemia**, **immune dysfunction**, and **infections**.

Key Points in Leukemia Development:

- Mutations in genes controlling **cell cycle regulation**.
- Faulty **checkpoints**, allowing defective cells to continue dividing.
- Failure in **apoptosis** (programmed cell death), which normally eliminates damaged cells.

Treatment Approaches:

- **Chemotherapy:** Uses drugs to target rapidly dividing cells.
- **Bone Marrow Transplant:** Replaces defective blood-forming cells with healthy donor cells.
- **Targeted Therapy:** Drugs that block specific cancer-related pathways.

Key Takeaways

- **Cell reproduction** is essential for **growth**, **repair**, and **reproduction** in living organisms.
- Processes like **mitosis** and **meiosis** ensure proper distribution of genetic material.
- **Cytokinesis** finalizes the physical separation into two cells.
- Errors in **cell cycle regulation** can lead to diseases like **leukemia**.

26. What is Genetic Material?

The **genetic material** of a cell contains the instructions for life. It directs how organisms develop, grow, function, and reproduce. In all known living organisms, this genetic material is composed of either **DNA** (deoxyribonucleic acid) or **RNA** (ribonucleic acid), depending on the organism. Genetic material is passed from **one generation to the next**, ensuring that species maintain continuity while allowing for genetic variation that drives **evolution**.

26.1. DNA – The Blueprint of Life

DNA is the primary genetic material in most living organisms, from bacteria to humans. It consists of two strands twisted into a **double helix**. Each strand is made up of a sequence of **nucleotides**, which are the building blocks of DNA. Each nucleotide contains:

- A **sugar** (deoxyribose)
- A **phosphate group**
- A **nitrogenous base** (adenine, thymine, cytosine, or guanine)

The sequence of these bases forms **genes**, which are the functional units of heredity.

26.2. RNA – The Messenger and More

RNA plays multiple roles in cells, acting as a **messenger**, a **catalyst**, and even a regulator of gene expression. Unlike DNA, **RNA**:

- Is usually **single-stranded**.
- Contains the sugar **ribose** instead of deoxyribose.
- Uses **uracil (U)** instead of thymine (T).

The main types of RNA include:

- **mRNA** (messenger RNA), which carries genetic information from DNA to the ribosome for **protein synthesis**.
- **tRNA** (transfer RNA), which helps incorporate **amino acids** into proteins.
- **rRNA** (ribosomal RNA), a structural component of ribosomes.

26.3. Genes – Instructions for Building Life

Genes are sequences of **DNA nucleotides** that code for specific proteins or functional RNA molecules. In humans, genes make up only a small portion of the **genome**, but they are responsible for virtually every biological process.

- Each gene contains **coding regions** (exons) and **non-coding regions** (introns).
- Genes are regulated by **promoters**, **enhancers**, and **silencers**, which determine when and how much a gene is expressed.

The complete set of genes in an organism is called its **genome**.

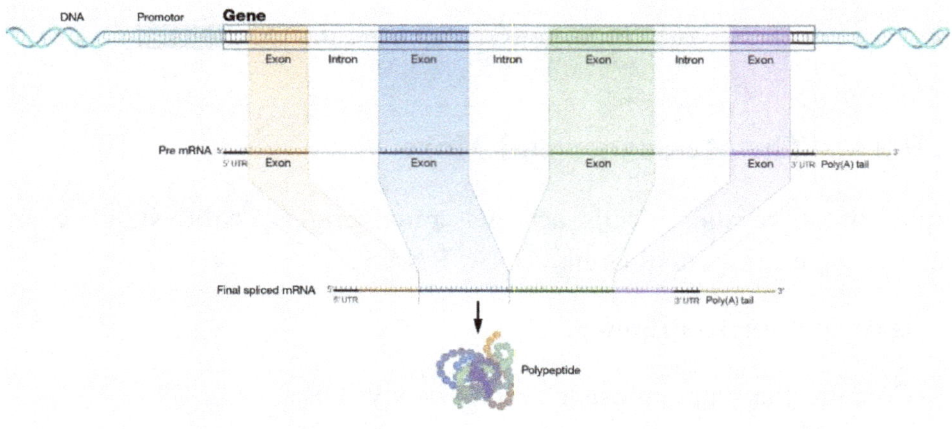

Figure 26.1: A gene structure showing exons, introns, and regulatory regions. Source: https://www.genome.gov/genetics-glossary/Gene

26.4. Chromosomes – Packaged DNA

In **eukaryotic cells**, **DNA** is organized into **chromosomes**. Each chromosome contains a long **DNA molecule** tightly wrapped around **histone proteins**. This structure allows:

- Efficient **storage** of genetic material.
- Proper **segregation** of DNA during cell division.
- Regulation of **gene expression** through changes in chromatin structure.

Humans have **46 chromosomes** (23 pairs), including **22 pairs of autosomes** and **one pair of sex chromosomes**.

Karyotype

A **karyotype** is a complete visual profile of an individual's **chromosomes**, typically displayed as an ordered arrangement of chromosome pairs, from the largest to the smallest, along with the **sex chromosomes**. This organized snapshot is obtained by **staining** and photographing **chromosomes** during cell division, usually in **metaphase**, when they are most condensed and visible.

Karyotyping is a valuable tool in **genetics** and **medical diagnostics**, as it allows scientists and doctors to detect **chromosomal abnormalities**, such as **extra chromosomes** (like in **Down syndrome**, which results from an extra copy of chromosome 21), **missing chromosomes**, or **structural changes** such as deletions, duplications, or translocations. A typical **human karyotype** (see Figure 26.2) includes **46 chromosomes**, organized into **22 pairs of autosomes** and **one pair of sex chromosomes** (XX for females, XY for males).

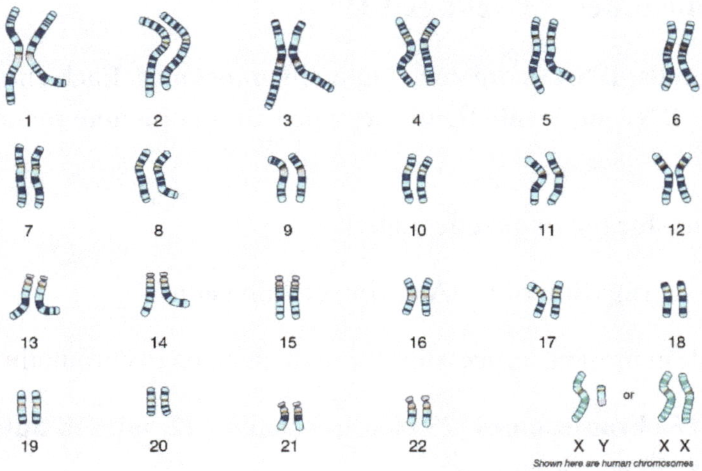

Figure 26.2: Human chromosome karyotype. Source: https://www.genome.gov/genetics-glossary/Karyotype

26.5. Non-coding DNA – More Than Just Junk

While **only about 1-2%** of the human genome codes for **proteins**, the rest consists of **non-coding DNA**. For decades, scientists considered this "junk DNA," but we now know it plays **crucial roles** in:

- Regulating **gene expression**.
- Controlling **chromosome structure**.
- Hosting sequences like **transposons** and **repetitive elements**.

Some **non-coding DNA** acts as **scaffolding** for chromosome organization, while other parts encode regulatory **RNA molecules**.

Chapter Summary

- **DNA** and **RNA** are the core **genetic materials** in cells.
- **Genes** encode **proteins** and functional RNAs.

- **Chromosomes** organize DNA for cell division and gene regulation.
- **Non-coding DNA** performs many vital **regulatory functions**.

27. The Replication of DNA

DNA replication is one of the most fundamental and essential processes in all living organisms. It is the mechanism by which a cell copies its **genetic material** to ensure that each new **daughter cell** receives a complete and accurate copy of the **DNA** after **cell division**. This process happens before a cell divides, either during **mitosis** or **meiosis**, and is tightly regulated to prevent errors that could lead to mutations or diseases.

27.1. Base Pairing

At the heart of DNA replication is the principle of **complementary base pairing**. In **double-stranded DNA**, **adenine (A)** pairs with **thymine (T)**, and **cytosine (C)** pairs with **guanine (G)**. When DNA unwinds during replication, each strand serves as a **template** for the formation of a new complementary strand. This ensures that the new DNA molecules are **identical** to the original. The **hydrogen bonds** between complementary bases allow the strands to be easily separated and rejoined during the replication process.

27.2. Linear Replication

In **eukaryotic cells**, DNA replication occurs along **linear chromosomes** within the **nucleus**. The replication process begins at **multiple origins of replication**, where the double helix opens into a **replication bubble**. **Replication forks** form at either end of the bubble, where specialized **enzymes** such as **DNA helicase** unzip the DNA strands.

As the forks progress, **DNA polymerase** synthesizes new complementary strands by adding **nucleotides** that match the template strand. Because DNA polymerase can only add nucleotides in the **5' to 3' direction**, replication occurs **continuously** on the **leading strand**, but **discontinuously** on the **lagging strand**, forming **Okazaki fragments** that are later joined by **DNA ligase**.

27.3. Self-replication

One of the most remarkable properties of DNA is its ability to **self-replicate** with high **fidelity**. Each strand of the **double helix** contains all the information necessary to construct a **complementary strand**, making DNA **semi-conservative** — meaning each newly synthesized DNA molecule consists of **one old (template) strand** and **one new strand**. This feature ensures that genetic information is passed **accurately** from one generation of cells to the next.

The **precision** of DNA replication is further supported by **proofreading enzymes**, which detect and correct most **errors** introduced during nucleotide addition. Despite this, occasional errors, known as **mutations**, can arise and contribute to **genetic variation** — a crucial driver of **evolution**. This highly **coordinated** process of DNA replication is **vital** for the survival, growth, and reproduction of all living cells, ensuring the faithful **transmission** of genetic information from one cell generation to the next.

28. What is Cell Reproduction?

Cell reproduction is the process by which cells generate **new cells**, enabling organisms to grow, repair damaged tissues, and reproduce. At its core, cell reproduction ensures that **genetic material** is faithfully copied and passed down from **parent cells** to **daughter cells**. This process is critical for the survival of both **unicellular** and **multicellular** organisms. Depending on the organism and the purpose of reproduction, cells can divide either by **mitosis** (for growth and repair) or **meiosis** (for producing gametes used in sexual reproduction).

28.1. Basic Steps in Cell Reproduction

Cell reproduction generally follows a **series of steps**, including:

1. **DNA replication**, ensuring that each daughter cell receives a complete copy of genetic material.
2. **Organelle duplication**, where structures like **mitochondria** and **centrosomes** are copied.
3. **Cell division**, splitting the parent cell into two or more daughter cells.
4. **Cytokinesis**, where the **cytoplasm** is divided between daughter cells.
5. **Cell differentiation** (in multicellular organisms), ensuring specialized functions.

28.2. Cell Division

Cell division is the central event of cell reproduction. It occurs in two main forms:

- **Mitosis**, where a parent cell divides to produce **two genetically identical daughter cells**, used for **growth**, **repair**, and **asexual reproduction**.
- **Meiosis**, where a parent cell divides to produce **four genetically distinct**

gametes, each with **half** the genetic material of the parent cell. This is essential for **sexual reproduction**.

28.3. Cytokinesis

Cytokinesis is the final stage of **cell division**, where the **cytoplasm** and **cell membrane** divide, physically separating the **daughter cells**. In **animal cells**, a **contractile ring** of **actin filaments** pinches the cell in two. In **plant cells**, a **cell plate** forms between daughter cells, eventually becoming the **new cell wall**.

28.4. Gamete Formation

Gametes — the **sperm** and **egg cells** — are formed through **meiosis**. These specialized cells are **haploid**, meaning they carry only **one set of chromosomes**. When gametes from two parents fuse during **fertilization**, they restore the **diploid** chromosome number in the resulting **zygote**, combining **genetic material** from both parents.

28.5. Chemical Replication

Beyond cellular mechanisms, **chemical replication** refers to the molecular processes by which **DNA** is **copied**, including **base pairing**, **polymerization**, and **error correction**. This chemical precision is critical for maintaining **genetic stability** across generations.

28.6. Leukemia Case Study

A clear example of **cell reproduction gone wrong** is **leukemia**, a type of **cancer** that affects **blood-forming tissues** such as **bone marrow**. In leukemia, the normal regulatory mechanisms controlling **cell division** fail, causing **uncontrolled proliferation** of abnormal **white blood cells**. These defective cells do not function properly and crowd out healthy cells, compromising the immune system and disrupting normal blood function.

Key takeaways from leukemia research:

- Mutations in genes controlling **cell cycle checkpoints** can trigger the disease.
- Some forms of leukemia are linked to **chromosomal translocations**, where segments of DNA break and rearrange, activating **oncogenes**.
- Modern treatments, such as **targeted therapies** and **bone marrow transplants**, focus on correcting or bypassing these cellular errors.

Cell reproduction, though often seamless, is a **highly regulated** process vital to **life** and **health**. Studying it helps researchers understand **development**, **disease**, and the evolution of complex life forms.

29. The Cell Cycle and Mitosis

The cell cycle is the series of events a **cell** undergoes to **grow**, **replicate its DNA**, and **divide** into two daughter cells. This carefully regulated process ensures the accurate duplication and distribution of **genetic material** and cellular components. Each stage of the **cell cycle** plays a vital role in maintaining the **health** and **function** of the organism, particularly in **multicellular** life forms.

29.1. Cell Cycle

The **cell cycle** is divided into **interphase** (where the cell grows and prepares for division) and **mitotic phase** (where the cell physically divides).

- **G1 Phase** – The cell grows and produces necessary proteins and organelles.
- **S Phase** – DNA replication occurs, ensuring that two complete **copies** of genetic material are present.
- **G2 Phase** – Final preparations for **mitosis** take place, including further growth and production of division-specific proteins.
- **M Phase** – The cell undergoes **mitosis**, dividing the nucleus, followed by **cytokinesis**, dividing the cytoplasm into two cells.

29.2. Checkpoints

Throughout the **cell cycle**, cells pass through several **checkpoints** that ensure all processes are proceeding correctly. These checkpoints act as **quality control stations**, assessing the **integrity** of the DNA and the readiness of the cell to divide.

- **G1 Checkpoint** – Ensures the cell is large enough and has enough **resources** to proceed.
- **G2 Checkpoint** – Confirms that **DNA replication** is complete and

accurate.

- **M Checkpoint** (Spindle Checkpoint) – Ensures that **chromosomes** are properly aligned and attached to **spindle fibers** before division.

Failures at these checkpoints can lead to **cancer** or other **genetic disorders**.

29.3. Negative Regulators

Negative regulators are proteins that **slow down** or **halt** the cell cycle if errors are detected. Some of the most important negative regulators include:

- **p53**, which can stop the cycle in response to **DNA damage** and initiate **apoptosis** (programmed cell death) if the damage is irreparable.
- **Rb (retinoblastoma protein)**, which prevents cells from advancing from **G1 to S phase** without proper growth signals.

Negative regulators are crucial for **preventing cancer**, as they help prevent **uncontrolled cell division**.

29.4. Telophase

Telophase is the **final stage** of mitosis, where **chromosomes** that were previously separated into opposite poles of the cell begin to decondense. During telophase:

- New **nuclear envelopes** form around each set of chromosomes.
- The **spindle apparatus** disassembles.
- The cell prepares for **cytokinesis**, which will split the **cytoplasm** into two.

Telophase ensures that each new cell receives a **complete nucleus**.

29.5. Anaphase

Anaphase is a **critical phase** of mitosis where **sister chromatids** (identical copies of a chromosome) are separated and pulled toward **opposite poles** of the cell. This separation ensures that each daughter cell receives a **complete set of chromosomes**.

- The **spindle fibers** shorten, pulling the chromatids apart.
- This movement is powered by **motor proteins** associated with the **spindle apparatus**.

Anaphase is one of the most **dynamic and visually striking stages** of mitosis.

The **cell cycle** and **mitosis** are fundamental to **growth**, **repair**, and **reproduction** in all living things. The seamless orchestration of these processes ensures **genetic fidelity**, proper tissue formation, and the **long-term stability** of organisms. Understanding these processes helps scientists **combat cancer**, study **stem cells**, and explore potential **regenerative medicine therapies**.

30. Cell Communities

Cells rarely exist in **isolation**. In **multicellular organisms** and **microbial colonies**, cells form **communities** where they **interact**, **communicate**, and **collaborate** to maintain **homeostasis**, ensure **survival**, and respond to **environmental changes**. These **cellular communities** are fundamental to the **functioning of tissues, organs, and entire ecosystems**.

30.1. How Do Cell Communities Operate?

Cell communities operate through **coordinated communication**, **chemical signaling**, and **physical interactions**. In tissues, for example, **epithelial cells**, **muscle cells**, and **nerve cells** work together to form **functional units**. In microbial biofilms, cells secrete substances that form **protective matrices** and regulate community **behaviors**, including **growth rates** and **nutrient sharing**.

In all cases, cell communities must:

- Share **resources**.
- Respond to **external cues**.
- Coordinate **growth**, **division**, and **specialized tasks**.
- Protect themselves from **threats**.

30.2. Bioluminescence

Some **cell communities**, particularly **marine microorganisms** like *Vibrio fischeri*, use **bioluminescence** as a form of **communication**. These bacteria emit **light** only when they reach a **critical population density**, a process called **quorum sensing**. This light production may help attract **hosts**, warn **predators**, or coordinate group behaviors.

30.3. Communication Between Cells

Cells within a community constantly exchange **chemical signals** to regulate **behavior** and **function**. These signals can:

- Trigger **cell division**.
- Initiate **differentiation**.
- Recruit neighboring cells for **repair**.
- Warn others of **stress** or **infection**.

In higher organisms, **hormones** and **neurotransmitters** mediate long-range **cell communication**, while **growth factors** and **cytokines** coordinate local responses. In bacteria, **quorum sensing molecules** regulate collective actions, such as forming biofilms or releasing toxins.

30.4. Interactions Between Subpopulations of Cells

In multicellular organisms, different **cell types** (subpopulations) must coordinate their **functions** to form **functional tissues** and **organs**. Examples include:

- **Immune cells** working with **epithelial cells** to maintain barriers.
- **Fibroblasts** producing structural proteins to support other cell types.
- **Neurons** communicating with **glial cells** for nerve function.

In microbial communities, some species **produce enzymes** to digest food, while others **scavenge nutrients**, resulting in a **division of labor**.

30.5. Control of Cell-cell Distances

To maintain proper **function**, cells must regulate their **spacing** within tissues. Cells do this by:

- Producing **adhesion molecules** that control how strongly they stick to neighbors.

- Sending **repulsion signals** to prevent overcrowding.
- Releasing **growth factors** that regulate population density.

This spacing is especially critical in the **development** of tissues, where cells must organize into **precise structures**, such as **capillaries**, **alveoli**, and **neural networks**.

30.6. Response to Changing Environments

Cell communities must **adapt** to environmental changes to survive. This could include:

- Adjusting **gene expression** to respond to stress.
- Reorganizing into **protective biofilms** in harsh conditions.
- Activating **immune responses** to infection.
- Modifying **nutrient uptake** strategies during starvation.

These responses rely on both **individual cellular sensing** and **community-wide signaling**, ensuring the survival of both the **individual cell** and the **community as a whole**.

Cell communities, whether simple or complex, represent a **fascinating example** of how life organizes itself into **cooperative networks**. From **bacterial biofilms** to **human tissues**, the ability of cells to **communicate, collaborate, and respond to changes** underpins the success of **all living systems**.

31. Central Dogma

The **Central Dogma** of molecular biology describes how **genetic information** flows within a biological system. Proposed by **Francis Crick** in 1958, it states that genetic information passes from **DNA** to **RNA** to **protein**, but not in the reverse order (with very few exceptions, such as retroviruses). This **framework** forms the foundation for **understanding gene expression** and how **traits** emerge from **genetic instructions**.

31.1. Transfer of Information from Nucleic Acid to Nucleic Acid

The first step of the Central Dogma involves the **transfer of genetic information** from one **nucleic acid** molecule to another — specifically, from **DNA to RNA** in a process called **transcription**.

- During **transcription**, **RNA polymerase** reads the **template strand** of DNA and synthesizes a complementary **mRNA** strand.
- This **mRNA** serves as a **messenger**, carrying the **genetic code** from the **nucleus** to the **cytoplasm** (in eukaryotes), where it can be **translated** into **protein**.

This step ensures the **faithful transfer** of **genetic instructions**, while allowing cells to **regulate gene expression** based on environmental signals.

31.2. Irreversibility of Translation

A **key principle** of the Central Dogma is that **translation** — the process of converting an **mRNA sequence** into a **protein sequence** — is **irreversible** under normal biological conditions.

- Once an **amino acid sequence** has been assembled into a **protein**, there is **no biological mechanism** to reverse the process and reconstruct the **original mRNA sequence** from the protein.
- This **irreversibility** ensures that **genetic information** flows in a **one-way**

direction, preserving the **integrity** of the **genetic code** stored in **DNA**.

This concept highlights how **proteins** are the **final functional products** of the genetic code, performing tasks **dictated by the genes** without directly feeding information back into the genome.

31.3. Codons in the Central Dogma

The **codon system** is central to the **Central Dogma**, translating **nucleotide sequences** into **amino acids** during **protein synthesis**.

- Each group of **three nucleotides** (a **codon**) corresponds to a specific **amino acid**.
- This **triplet code** is **universal** across almost all known organisms, emphasizing its **fundamental role** in life.
- Some codons act as **start** or **stop signals**, defining where **translation** begins and ends.

The existence of **64 codons**, encoding **20 amino acids** plus control signals, ensures that the **genetic code** is both **specific** and **redundant**, providing some tolerance for **mutations**.

31.4. Mechanisms that are Physically Feasible to Realize the Central Dogma

The Central Dogma is not just a **conceptual framework** — it is realized through **well-characterized molecular mechanisms**. These processes are **physically feasible** because they rely on:

- **Base pairing rules** (A pairs with T/U, C pairs with G) to ensure **accurate copying** of **nucleic acids**.
- **Enzymes** such as **RNA polymerase** and **ribosomes**, which act as **molecular machines**, carrying out the **synthesis** of RNA and proteins.

- **Thermodynamic stability** of **double-stranded DNA**, which protects genetic information while allowing controlled **separation** for **transcription** and **replication**.
- **Adaptor molecules** like **tRNA**, which physically link **codons** to **amino acids**, ensuring the correct **sequence assembly** during **translation**.

These mechanisms form a **seamless chain**, enabling life to **store, copy, and express genetic information** with remarkable **fidelity**.

32. How Genes Make Proteins

The **process of protein synthesis** — how **genes make proteins** — is central to the **expression of genetic information**. This process, called **gene expression**, follows the **Central Dogma** covered in the previous chapter, moving from **DNA** to **RNA** to **protein**. Genes contain **coding sequences** that provide the **instructions** for building proteins, and **non-coding regions** that regulate when and how much protein is made. This **highly regulated process** allows cells to respond to changes in their environment, differentiate into specialized types, and evolve over time.

32.1. Non-coding or RNA Genes

Not all genes encode proteins. Some genes produce **functional RNA molecules** instead, including:

- **Transfer RNA (tRNA)** — assists in **translation** by carrying **amino acids** to the **ribosome**.
- **Ribosomal RNA (rRNA)** — forms the core structural and **catalytic components** of ribosomes.
- **MicroRNA (miRNA)** — regulates gene expression by **binding to mRNA** and blocking **translation**.

These **non-coding RNAs** perform **essential cellular functions**, even though they are not translated into **proteins**.

32.2. Transfer RNA (tRNA)

tRNA is a **small RNA molecule** that acts as an **adapter** during **translation**.

- Each **tRNA** has an **anticodon** that pairs with a specific **codon** on the **mRNA**.
- On the opposite end, the tRNA carries the **corresponding amino acid**,

ensuring that the **correct amino acid** is added to the growing **polypeptide chain**.

The **accurate matching** of tRNA to codons ensures **proper protein assembly**.

32.3. Codons

Codons are **three-nucleotide sequences** in **mRNA** that specify individual **amino acids** or **regulatory signals** (start and stop codons).

- There are **64 codons**, with most specifying **amino acids**.
- **AUG** is the **start codon**, and **UAA, UAG, and UGA** are the **stop codons**.

The **order of codons** in a gene determines the **sequence of amino acids** in the final **protein**.

32.4. mRNA

Messenger RNA (mRNA) is the **intermediate copy** of a gene that is directly used to **build proteins**.

- During **transcription**, the **DNA sequence** of a gene is copied into **mRNA**.
- This **mRNA** leaves the nucleus (in **eukaryotes**) and travels to the **ribosome**, where it directs **protein synthesis**.
- **mRNA sequences** are read in groups of **three nucleotides** (codons), each specifying a particular **amino acid**.

The **lifespan** and **stability** of mRNA molecules help control **gene expression**.

32.5. tRNA

As covered above, **tRNA** plays a **vital role** in the translation process. Each tRNA type is **specific** for one **amino acid**, ensuring the **correct incorporation** into the protein.

- The **tRNA anticodon** binds to the **mRNA codon**, ensuring correct pairing.
- The **ribosome** links the incoming **amino acids** to form a **polypeptide chain**.

This **stepwise process** ensures **high accuracy** in protein synthesis.

32.6. Cell's Origin

The **mechanisms for gene expression** — from **transcription** to **translation** — are so **universally conserved** that they provide important **clues** about the **origin of cells**.

- **The genetic code** is nearly **identical** across all known life forms, suggesting that all life shares a **common ancestor**.
- The **basic machinery** (RNA polymerase, ribosomes, etc.) has been retained with only **minor variations**.

This **conservation** underscores the **central importance** of **accurate gene expression** in the **survival** and **evolution** of life.

32.7. Evolution of Gene Regulation

While the **basic process** of gene expression is **conserved**, **how genes are regulated** has evolved dramatically.

- Early organisms may have relied on **simple feedback loops** to regulate genes.
- More **complex organisms** evolved **sophisticated regulatory networks** involving **transcription factors**, **enhancers**, and **non-coding RNAs**.
- These **evolved mechanisms** allow organisms to **fine-tune gene expression** in response to **developmental cues**, **environmental changes**, and **stress**.

Understanding how **genes make proteins** — and how cells control **when and how much protein** they make — is essential to understanding **life itself**, from **basic biology** to **genetic diseases** and **biotechnology**.

33. DNA Repair and Recombination

DNA Repair and **recombination** are essential processes that **maintain genomic stability**, **correct damage**, and **enable genetic diversity**. Every day, **cells are exposed** to internal and external sources of **DNA damage**, including **UV radiation**, **chemical mutagens**, and **normal metabolic activities**. Without efficient **repair mechanisms**, **mutations** would accumulate, leading to **diseases** such as **cancer**.

Recombination, on the other hand, allows for the **exchange of genetic material** between DNA molecules, playing a key role in **meiosis** and **genetic variation**. Both processes rely on a suite of **specialized proteins** to detect, process, and repair **damaged DNA**.

33.1. Essential Proteins for DNA Repair and Recombination

A variety of **proteins** ensure that **DNA integrity** is preserved. These **proteins** detect damage, signal for repair, and physically correct the **DNA lesion**. Some of the most important include:

- **XPF/ERCC1** — Involved in **nucleotide excision repair** (NER), this **endonuclease complex** cuts the **damaged strand** of DNA, allowing for **removal** of faulty segments.

- **XPG** — Another critical **endonuclease** involved in **NER**, **XPG** cuts on the opposite side of the damage compared to **XPF/ERCC1**, ensuring complete **removal** of the **damaged section**.

- **XRCC2** — Plays a key role in **homologous recombination**, particularly in **repairing double-strand breaks**. It helps **stabilize the repair complex** and promotes **error-free repair**.

- **XPA** — Functions as a **damage verification protein**, ensuring that the correct repair **pathway** is activated. It binds directly to **damaged DNA sites**.

- **XPD** — A **helicase** that unwinds the DNA helix at sites of damage, allowing other **repair factors** to access and process the lesion.

33.2. Excision Repair

Excision repair is one of the most versatile and **important DNA repair mechanisms**. It corrects:

- **Damage caused by UV light** (thymine dimers)
- **Bulky adducts caused by chemical exposure**
- **Oxidative damage from cellular metabolism**

The process involves several coordinated steps:

1. **Damage Recognition** — Specialized proteins (like **XPA**) scan the DNA and recognize **abnormal structures**.
2. **Unwinding and Verification** — Proteins like **XPD** help unwind the DNA around the damage.
3. **Excision** — Endonucleases such as **XPF/ERCC1** and **XPG** cut out the damaged section.
4. **Resynthesis** — DNA polymerases **synthesize new DNA** to replace the removed section.
5. **Ligation** — DNA ligase **seals the final gap**, restoring the **DNA strand**.

33.3. XPF/ERCC1

The **XPF/ERCC1 complex** acts as a **structure-specific endonuclease**, recognizing the **junction between single- and double-stranded DNA** and cutting at precise **locations**. It is critical for:

- **Nucleotide excision repair** (NER)

- **Repair of interstrand crosslinks** (damage that links two DNA strands together)
- **Some forms of recombination repair**

33.4. XPG

XPG functions similarly to **XPF/ERCC1**, but it **cuts on the opposite side** of the **damaged site**. It works closely with other **repair factors** and is essential for **NER** to complete. **Mutations in XPG** can lead to **xeroderma pigmentosum**, a disorder characterized by extreme sensitivity to UV light and increased cancer risk.

33.5. XRCC2

XRCC2 is essential for **homologous recombination** repair (HRR), a pathway that **accurately repairs double-strand breaks**. It:

- Forms a **complex with other RAD proteins**.
- Ensures correct **pairing and exchange** between **homologous DNA sequences**.
- Helps cells **survive DNA damage** caused by **ionizing radiation** and **chemotherapy drugs**.

33.6. XPA

XPA acts as a **scaffold protein** in **nucleotide excision repair**, helping:

- Confirm the **presence** of damage.
- Recruit other repair **factors** to the site.
- Coordinate the **cutting and replacement** steps.

Mutations in **XPA** also lead to **xeroderma pigmentosum**, showing how crucial this protein is for **UV protection**.

33.7. XPD

XPD is a component of the **TFIIH complex**, which plays dual roles in:

- **Transcription initiation**.
- **Nucleotide excision repair**.

As a **helicase**, **XPD** unwinds the **DNA helix**, creating an **open bubble** where **endonucleases** like **XPF/ERCC1** and **XPG** can act.

Summary

DNA repair and recombination processes ensure **genetic stability** while also contributing to **genetic diversity**. Without these crucial mechanisms, **cells would accumulate mutations**, increasing the risk of **cancer**, **genetic disorders**, and **cell death**. By understanding the **key proteins** and processes involved, researchers are developing better **therapies** for **DNA repair deficiencies** and even exploring ways to enhance **DNA repair capacity** to fight **aging** and **disease**.

34. Gene Regulation

Gene regulation is the complex process by which **cells control the expression** of their **genes** — deciding when, where, and how much of a **gene product** (usually a **protein** or **functional RNA**) is made. This regulation is essential for ensuring that cells **respond appropriately** to their **environment**, **differentiate into specialized cells**, and **conserve energy** by only producing what is necessary.

In both **prokaryotic** and **eukaryotic cells**, gene regulation is highly **coordinated** and **multilayered**, involving **many regulatory elements**, **proteins**, and **signaling pathways**.

34.1. Levels of Gene Regulation

Gene regulation can occur at **multiple stages**, including:

- **Transcriptional Regulation** – Controlling whether a gene is **transcribed into RNA**. This is often the **most important and energy-efficient level** of control.
- **Post-transcriptional Regulation** – Modifying or controlling the **mRNA** after it is transcribed, such as through **splicing**, **editing**, **transport**, and **stability**.
- **Translational Regulation** – Controlling how efficiently an **mRNA** is **translated into a protein**.
- **Post-translational Regulation** – Modifying the **protein** after translation, such as by **folding**, **cleaving**, **activating**, or **degrading** it.

34.2. Transcriptional Regulation in Prokaryotes

Prokaryotes, such as **bacteria**, use simple but highly effective **gene regulation systems**. The **operon model**, discovered in **E. coli**, is a classic example.

The Operon System

An **operon** is a **cluster of genes** controlled by a **single promoter** and regulated together. For example:

- **Lac operon** – Controls genes involved in **lactose metabolism**.
- **Trp operon** – Controls genes involved in **tryptophan synthesis**.

In these systems, **regulatory proteins** bind to **DNA sequences** called **operators**, either **blocking** or **allowing transcription** depending on the presence of **specific molecules** like lactose or tryptophan.

34.3 Gene Regulation in Eukaryotes

Eukaryotic gene regulation is far more **complex** due to:

- Presence of a **nucleus**, requiring transport between nucleus and cytoplasm.
- More extensive **chromatin structure**, which affects gene accessibility.
- A larger genome with more regulatory **DNA sequences**.

Key Elements of Eukaryotic Gene Regulation

- **Enhancers and Silencers** – DNA sequences that increase or decrease gene expression.
- **Transcription Factors** – Proteins that bind to specific **DNA sequences** to **activate** or **repress** transcription.
- **Epigenetic Modifications** – Chemical changes to **DNA** (like **methylation**) or to **histones** (like **acetylation**) that alter gene expression without changing the **DNA sequence** itself.
- **RNA Interference (RNAi)** – Small RNAs, such as **miRNAs**, that can bind to **mRNA** and **block translation** or trigger **mRNA degradation**.

34.4. Environmental and Developmental Regulation

Cells do not regulate genes **randomly** — gene expression changes in response to:

- **Environmental stimuli** – Light, temperature, chemicals, and stress.
- **Internal signals** – Hormones, **growth factors**, and nutrients.
- **Developmental programs** – Genes are turned on and off during **embryo development**, **cell differentiation**, and **aging**.

34.5. Examples of Gene Regulation in Action

- **Heat shock response** – When cells experience high **temperatures**, they activate **heat shock genes**, which produce **proteins** that help **refold damaged proteins**.
- **Hormonal regulation** – In **humans**, the **insulin gene** is tightly regulated based on **blood sugar levels**.
- **Immune response** – Infection triggers rapid activation of **genes coding for cytokines** and **antibodies**.

Summary

Gene regulation ensures that cells:

- **Use resources efficiently**.
- **Respond appropriately** to changes.
- **Differentiate into specialized cell types**.
- **Avoid unnecessary protein production**.
- **Maintain genomic stability and cellular health**.

Whether in simple **bacteria** or **complex multicellular organisms**, **gene**

regulation is at the heart of **biological control systems**, enabling life to adapt, survive, and thrive in constantly changing environments.

35. Genetic Engineering of Plants

Genetic engineering of plants is the process of **modifying the genetic material** of plants to achieve **desired traits**. By introducing, removing, or altering **genes**, scientists can create plants with **improved yields**, **resistance to pests**, **drought tolerance**, and **enhanced nutritional value**. This field is part of a broader area called **agricultural biotechnology** and has had a **profound impact** on global food production, sustainability, and environmental protection.

35.1. Techniques Used in Plant Genetic Engineering

The following methods are commonly used to modify plant genomes:

- **Recombinant DNA Technology** – Scientists isolate **specific genes** from one organism (like bacteria or another plant) and insert them into a plant's genome using **plasmids** or **vectors**.
- **Gene Gun Method** – Small particles coated with **DNA** are shot directly into **plant cells**, delivering new genetic material.
- **Agrobacterium-mediated Transformation** – This method uses the natural ability of the **Agrobacterium tumefaciens** bacterium to transfer **genes** into plant cells.
- **CRISPR-Cas9 Gene Editing** – This modern technique allows for **precise editing** of a plant's existing **genes** to turn them on or off or to replace defective sequences.

35.2. Traits Introduced through Genetic Engineering

The most common genetically engineered traits include:

- **Pest Resistance** – Plants like **Bt cotton** and **Bt corn** produce their own **insecticidal proteins** to ward off pests.
- **Herbicide Tolerance** – Crops like **Roundup Ready soybeans** can

survive being sprayed with certain **herbicides**, simplifying weed control.

- **Drought and Salinity Tolerance** – Genetic modifications can enhance a plant's ability to withstand **environmental stress**, helping farmers in **arid regions**.
- **Nutritional Enhancement** – Crops like **Golden Rice** are engineered to produce higher levels of **vitamin A** to combat malnutrition.
- **Disease Resistance** – Some genetically engineered plants resist **viral, bacterial, or fungal infections**, reducing the need for **chemical treatments**.

35.3. Benefits of Genetically Engineered Plants

- **Higher Yields** – Genetic improvements contribute to **increased crop production**, ensuring **food security**.
- **Reduced Chemical Use** – Plants engineered for **pest and disease resistance** require fewer **pesticides**, lowering environmental pollution.
- **Adaptation to Climate Change** – Drought- and heat-tolerant crops help **farmers adapt** to **changing weather patterns**.
- **Nutritional Enhancement** – Biofortified crops address **nutrient deficiencies** in vulnerable populations.
- **Economic Benefits** – Higher yields and reduced input costs improve **farmer incomes**.

35.4 Environmental and Ethical Considerations

While genetic engineering offers **numerous benefits**, it also raises **environmental and ethical concerns**, such as:

- **Cross-pollination** – Genes from genetically modified (GM) plants can spread to **wild relatives**, potentially altering **natural ecosystems**.

- **Development of Resistance** – Pests or weeds may develop resistance to **engineered traits**, leading to **new agricultural challenges**.
- **Biodiversity Impact** – Large-scale planting of a few genetically modified crops may **reduce crop diversity**, increasing vulnerability to **disease outbreaks**.
- **Ethical Concerns** – Some groups oppose **genetic modification** based on **religious beliefs** or concerns about **corporate control** over seeds.

35.5. Regulatory Oversight

Due to these concerns, genetically engineered plants are **subject to strict regulation** in many countries. Regulatory agencies such as:

- **U.S. Department of Agriculture (USDA)**
- **Environmental Protection Agency (EPA)**
- **Food and Drug Administration (FDA)**

These agencies **evaluate the safety** of genetically engineered crops for **human consumption**, **environmental impact**, and **agricultural use** before approving them for **commercial production**.

35.6. Examples of Genetically Engineered Crops

Crop	Trait	Benefit
Golden Rice	Enhanced Vitamin A	Addresses **Vitamin A deficiency**
Bt Corn	Insect Resistance	Reduces **insecticide use**
Roundup Ready Soybeans	Herbicide Tolerance	Allows **simplified weed management**
Drought-Tolerant Maize	Drought Tolerance	Improves **yield under dry conditions**
Virus-Resistant Papaya	Disease Resistance	Saves **papaya industry from viral outbreaks**

Summary

Genetic engineering of plants is a **powerful tool** for improving **global agriculture**, offering solutions to **hunger, malnutrition, and environmental sustainability**. While it holds great promise, it also requires **responsible research**, **regulation**, and **public engagement** to address **concerns** and ensure that the benefits are shared **equitably**.

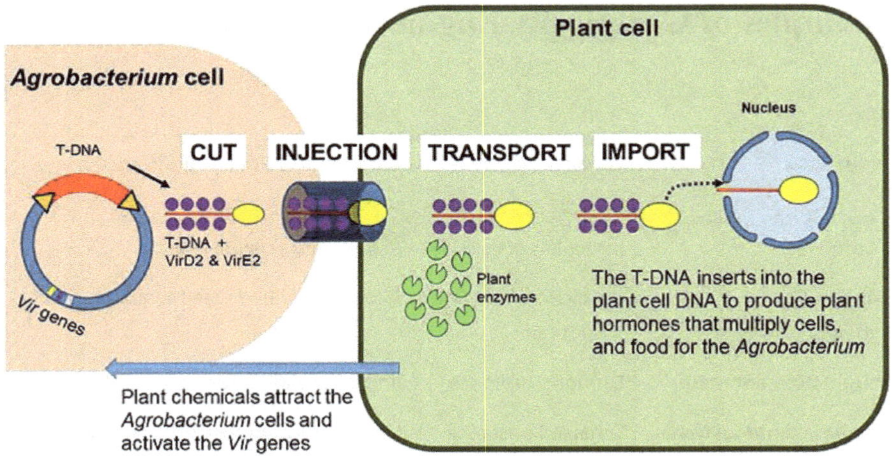

Figure 35.1: Process of inserting a gene into a plant cell using Agrobacterium tumefaciens. Source: https://kids.frontiersin.org/articles/10.3389/frym.2020.00064

36. Using Genetic Engineering in Animals and Humans

36.1. Genetic Engineering Technology

This technology allows scientists to **insert, remove, or modify genes** to produce desired traits. In both **animals** and **humans**, genetic engineering has the potential to **treat diseases, enhance physical traits, and even alter inherited characteristics**. Advances in technologies like **CRISPR-Cas9** have made gene editing faster, cheaper, and more precise than ever before.

36.2. Genetic Engineering in Animals

The genetic modification of animals serves many purposes, including:

- **Disease Research:** Animals, such as genetically modified mice, are used to study **human diseases** like cancer, diabetes, and neurological disorders. These animals are known as **transgenic models**.
- **Food Production:** Genetically engineered animals, such as **AquaBounty salmon**, are designed to grow faster and **require fewer resources**.
- **Pharmaceutical Production:** Animals can be genetically modified to produce **medicines** in their milk, eggs, or blood—a process known as **pharming**.
- **Conservation Efforts:** Scientists are exploring genetic engineering to **increase genetic diversity** in endangered species or even **resurrect extinct species** through **de-extinction projects**.

Example:

- **Glow-in-the-dark fish** (GloFish) have been genetically modified to **express fluorescent proteins**, originally from jellyfish or coral.

36.3. Human Genetic Engineering

Human genetic engineering involves the **modification of human DNA** to **treat or prevent disease**, enhance physical traits, or even alter **cognitive abilities**. There are two major categories:

- **Somatic Gene Therapy:** Genetic changes are made in **body cells** (not sperm or eggs) to correct a disease in an individual. These changes are **not passed on to offspring**.
- **Germline Genetic Engineering:** Modifications are made to **sperm, eggs, or embryos**, meaning the changes are **heritable**—they can be passed down to future generations. This type of genetic engineering is highly controversial and is largely prohibited in many countries.

Current and Potential Applications

- **Gene Therapy:** Correcting defective genes responsible for **diseases** such as cystic fibrosis, sickle cell anemia, and muscular dystrophy.
- **Prevention of Genetic Disorders:** Screening embryos for **genetic diseases** (preimplantation genetic diagnosis, PGD) is already used, but in the future, defective genes could be directly **corrected**.
- **Enhancement:** The possibility of using genetic engineering to **enhance intelligence, physical strength, or appearance**—sometimes referred to as creating **designer babies**—is a controversial ethical frontier.

Example:

- **CRISPR babies in China** (2018) – A scientist edited the genomes of twin girls to make them **resistant to HIV**, sparking a global ethical debate.

36.4. Gene Doping

Gene doping is the use of **genetic engineering techniques** to enhance **athletic**

performance by altering genes related to **muscle growth, endurance, or recovery**. Unlike traditional doping, which involves **taking drugs or hormones**, gene doping directly **modifies an athlete's DNA**.

Potential Targets for Gene Doping

- **Erythropoietin (EPO):** To increase **red blood cell production** for greater endurance.
- **Myostatin Inhibition:** To promote **muscle growth**.
- **VEGF (Vascular Endothelial Growth Factor):** To improve **oxygen delivery to muscles**.

Ethical and Health Concerns

- Gene doping is **banned** by organizations like the **World Anti-Doping Agency (WADA)**.
- The **uncontrolled use of gene editing** could have **unintended side effects**, such as **cancer**, **immune system complications**, or **unpredictable genetic mutations**.

36.5. Ethical and Societal Considerations

The ability to edit genes in animals and humans comes with significant **ethical concerns**, including:

- **Consent:** Is it ethical to modify the genome of **future generations** who cannot consent?
- **Equity:** Could genetic enhancement create **a genetic underclass**—dividing society into **genetically enhanced** and **natural** populations?
- **Safety:** The long-term effects of **gene editing**, especially **germline edits**, are largely unknown.

- **Biodiversity Impact:** Genetically modified animals could **disrupt ecosystems** if released into the wild.

Summary

Genetic engineering is **revolutionizing medicine, agriculture, and biotechnology**, but its use in **animals and humans** requires **caution, ethical reflection, and strict oversight**. As research advances, society will need to **balance innovation with responsibility**, ensuring that genetic technologies are used to **benefit humanity while minimizing harm**.

37. What is Gene Therapy?

37.1. Introduction to Gene Therapy

Gene therapy is a revolutionary **medical technique** that **treats or prevents disease** by directly **modifying the genetic material** inside a patient's cells. Unlike conventional drugs that only treat symptoms, gene therapy addresses **the root cause** of genetic disorders by either **replacing, repairing, or regulating defective genes**. This emerging field holds tremendous potential for curing diseases that were once considered **incurable**.

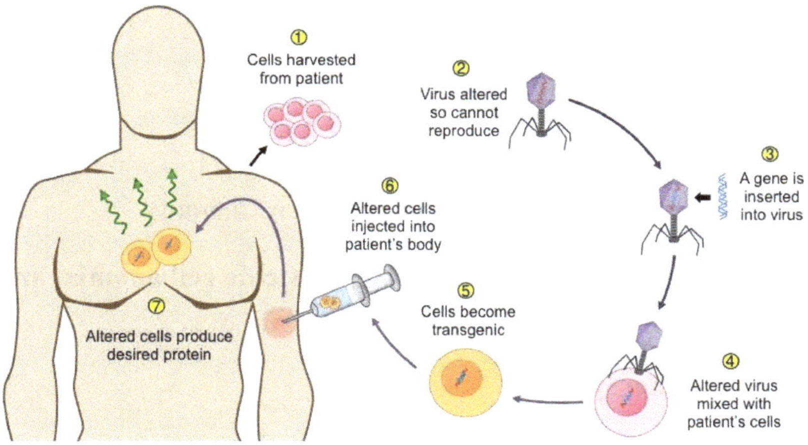

Figure 37.1: Overview of Gene Therapy Process. Source: https://old-ib.bioninja.com.au/standard-level/topic-3-genetics/35-genetic-modification-and/gene-therapy.html

37.2. How Does Gene Therapy Work?

The core goal of gene therapy is to **correct faulty genes** responsible for disease. This can be achieved through several approaches:

- **Gene Replacement:** Introducing a healthy copy of a defective gene into cells to **restore normal function**.
- **Gene Editing:** Using technologies like **CRISPR-Cas9** to **cut, remove, or**

repair mutated genes.

- **Gene Silencing:** Inactivating **harmful or overactive genes** that contribute to disease.
- **Gene Addition:** Adding entirely new genes that help **combat disease**, such as genes that enhance the **immune response** to cancer.

37.3. Types of Gene Therapy

Gene therapy can target different types of cells, leading to two major approaches:

1. Somatic Gene Therapy

- Alters **non-reproductive cells** (body cells).
- Changes affect only the **treated individual**.
- Currently the most **widely used form** of gene therapy.
- Used to treat diseases like **cystic fibrosis, sickle cell anemia, and some cancers**.

2. Germline Gene Therapy

- Alters **reproductive cells** (sperm, eggs, or embryos).
- Changes are **inherited by future generations**.
- Highly **controversial** and **largely banned** in most countries due to **ethical concerns**.
- Potential applications include preventing **inherited genetic diseases** from being passed down.

37.4. Delivery Methods

For gene therapy to work, therapeutic genes must be **delivered into the right**

cells. Scientists use **vectors**, which act as delivery vehicles, to **transport genes into cells**.

Common Delivery Methods

- **Viral Vectors:** Modified viruses (like adenoviruses or lentiviruses) that carry therapeutic genes into human cells.
- **Non-viral Methods:** Techniques such as **naked DNA injection, liposomes**, or **nanoparticles**, which deliver genes without using viruses.

Each method has its **advantages and challenges** in terms of **efficiency, safety, and immune response**.

37.5. Examples of Gene Therapy Applications

Gene therapy is already being used or researched to treat:

- **Inherited Disorders:** Such as **severe combined immunodeficiency (SCID)**, **spinal muscular atrophy (SMA)**, and **hemophilia**.
- **Cancer:** Engineering immune cells (CAR-T therapy) to **recognize and attack tumors**.
- **Blindness:** Introducing working genes into retinal cells to **restore vision** in certain inherited forms of blindness.
- **Blood Disorders:** Correcting gene mutations in **sickle cell disease** and **beta-thalassemia**.

37.6. Benefits and Potential of Gene Therapy

- **Permanent Cure:** Gene therapy can potentially provide a **lifelong cure** after a single treatment.
- **Targeted Treatment:** Therapy is tailored to an individual's **unique genetic makeup**, increasing precision.

- **New Hope:** Offers treatment options for **rare genetic diseases** that currently have no cure.

37.7. Risks and Challenges

- **Immune Response:** The body may **reject the viral vector**, reducing effectiveness.
- **Off-target Effects:** Gene editing may accidentally **modify unintended parts of the genome**, causing **new mutations**.
- **High Costs:** Gene therapy treatments, like **Zolgensma for spinal muscular atrophy**, can cost **millions of dollars per patient**.
- **Ethical Concerns:** Germline editing raises questions about **designer babies** and altering **the human gene pool**.

37.8. Future of Gene Therapy

With **advances in genome sequencing, CRISPR technology**, and **personalized medicine**, gene therapy will become **safer, more accessible, and more effective**. Scientists are also exploring **epigenetic therapy**, which targets **gene expression** without altering the underlying **DNA sequence**.

Summary

Gene therapy is **transforming medicine** by addressing the **root causes of genetic diseases** at the DNA level. While it holds incredible promise for treating a wide range of conditions, **ethical considerations, safety, and accessibility** must be carefully managed as this technology advances.

38. Conclusion

Now that you have completed this book, I hope you will agree with me that Biology is not a difficult or painful subject to study — in fact, it can be fascinating and rewarding when approached with curiosity and the right foundation. Biology is a subject that continuously builds upon itself, meaning that mastering the basic concepts is essential before you can fully understand the more complex topics.

I also hope you feel confident in applying much of what you've learned throughout this book to your future studies and even your daily life. Understanding the **vocabulary**, key **concepts**, and how they relate to one another will greatly improve your comprehension and help you excel in **exams** and **practical applications**.

The best way to succeed in **biology** is to stay organized, keep up with the material regularly, and practice applying what you've learned.

On behalf of all the **educators** and **creators** at **AB Publisher LLC**, I wish you the very best in your studies and future career. If you have any **questions**, **feedback**, or **concerns**, feel free to reach out to us at the email address below.

Cheers,
Bolakale Aremu (Educator)
AB Publisher LLC
ABPublisherLLC@gmail.com

www.ingramcontent.com/pod-product-compliance
Lightning Source LLC
LaVergne TN
LVHW060038210926
848481LV00013B/3898

* 9 7 9 8 3 4 9 2 4 2 8 1 6 *